Sustainable Development Goals Series

World leaders adopted Sustainable Development Goals (SDGs) as part of the 2030 Agenda for Sustainable Development. Providing in-depth knowledge, this series fosters comprehensive research on these global targets to end poverty, fight inequality and injustice, and tackle climate change.

The sustainability of our planet is currently a major concern for the global community and has been a central theme for a number of major global initiatives in recent years. Perceiving a dire need for concrete benchmarks toward sustainable development, the United Nations and world leaders formulated the targets that make up the seventeen goals. The SDGs call for action by all countries to promote prosperity while protecting Earth and its life support systems. This series on the Sustainable Development Goals aims to provide a comprehensive platform for scientific, teaching and research communities working on various global issues in the field of geography, earth sciences, environmental science, social sciences, engineering, policy, planning, and human geosciences in order to contribute knowledge towards achieving the current 17 Sustainable Development Goals.

This Series is organized into eighteen subseries: one based around each of the seventeen Sustainable Development Goals, and an eighteenth subseries, "Connecting the Goals," which serves as a home for volumes addressing multiple goals or studying the SDGs as a whole. Each subseries is guided by an expert Subseries Advisor.

Contributions are welcome from scientists, policy makers and researchers working in fields related to any of the SDGs. If you are interested in contributing to the series, please contact the Publisher: Zachary Romano [Zachary.Romano@springer.com].

More information about this series at http://www.springer.com/series/15486

Ana Sabogal

Urban Ecology

A Case Study of Lima City, Perú

Springer

Ana Sabogal
Pontificia Universidad Católica del Perú
Lima, Peru

ISSN 2523-3084 ISSN 2523-3092 (electronic)
Sustainable Development Goals Series
ISBN 978-3-030-69904-8 ISBN 978-3-030-69905-5 (eBook)
https://doi.org/10.1007/978-3-030-69905-5

This Springer imprint is published by the registered company Springer Nature Switzerland AG
The registered company address is: Gewerbestrasse 11, 6330 Cham, Switzerland

I would like to dedicate my work to my children who are my inspiration.

Preface

Writing this book has taken time. Each book matures at its own pace. It usually takes at least a decade since the first moment a person thinks about a subject and the moment a book is finished; in this case it has taken even longer. As an agricultural engineer I always enjoyed planting and walking around parks. When I gazed at the plants, I was drawn by the poor conditions they had to grow in. I did not understand why plants were treated like simple objects in design without considering their life span and special requirements of soil, water and air. While teaching at the Diplomacy Program in Management, Design of Green Areas and Urban Arboriculture in the Universidad Nacional Agraria in La Molina, I had the chance to think about this subject together with my students for more than 10 years and I learnt a lot through their questions and concerns which mainly aimed for practical solutions to the problems.

The definition of urban ecology as a science that studies plants in the context of a city has a different view in which vegetation is found in a surrounding where human beings intervene. This new science does not take plants out of context, considering only form and colour without contemplating its needs and surroundings. Urban ecology studies nature from a new perspective where new contexts are created giving place to unknown ecosystems where disturbances and changes are constant and plants, animals, together with human beings strive to survive. This view does not take green areas out of context but also includes its connections. Progressively new spaces of study are introduced to Urban Ecology research. Aspects such as ecosystems which develop within built spaces become important both for ecology and human health. Spaces are renewed with green walls and roofs, rooftops, balconies and even areas inside the houses or apartments are important for urban ecology. It is also related to the unrestrained increase of population living in cities and food safety. It is necessary to raise the question of how we will be able to feed ourselves if we all live in the city. Urban bio-orchards are only possible if the vegetables we grow are not contaminated by the toxic gases released in the city. Solutions to these questions can be found in different types of sciences from engineering to social sciences, hence, Urban ecology is essentially an interdisciplinary science that should consider multiple factors both natural and anthropic. In the twenty-first century, more than 50% of the world's population lives in cities and so the study of Urban Ecosystem has become a huge necessity. This book focuses on the study for making the city livable and on the ecosystems of Lima. The chapter prioritizes the Sustainable Development Goal 11 and in this way is oriented to propose sustainable solutions for Lima, so that Lima becomes a more sustainable, inclusive, safe and resilient city.

This book does a study that goes from landscape ecology theory to agronomic conditions required by plants and botanical species selection in context, with examples focused mainly on Peru and in the case of Urban Ecology on Lima city. In terms of green areas, Lima confronts a challenge that few other cities need to face due to lack of resources and water. When we revalue spaces that have water such as the ocean cliffs or the river Rimac banks, we rethink the subject from different perspectives. The big challenge of how to create green spaces in the rest of the city remains as well as how to eliminate inequality in the access to these areas.

Berlin, Germany Ana Sabogal
2020

Acknowledgements I wish to thank the Pontificia Universidad Católica del Perú for giving me a research semester to finish the writing process of this book and the *Programa trAndeS* for the economic support that gave me time to think and write and allowed me to systematize the information for this book. For the traslatin of the book I wish thank Katherine Flores.

Contents

About the Author

Ana Sabogal is an Agronomic Engineer, Doctor in natural sciences from the Technic University of Berlin, Plant Ecologist and Academic Scholar. Her current research includes the study of the impact of vegetation changes and the human impact on the Amazonian forest ecosystem. Her publications are about plant ecology of moorland ecosystems, plant distribution, in degraded grassing ecosystems and the influence of human impact. Other important previous research and publications included study of vegetation, grazing in the forests of northern Peru with emphasis on the distribution of *Ipomoea Carnea* Jacq. and management of dry forests in the northern coast of Peru. She developed and established the postgraduate program of Master study in Environmental Development at the Pontificia Universidad Católica del Peru and was previously Director of Research and Information in the Peruvian Ministry of Environment (2012–2013). Her work as a researcher and consultant in an International and national level is oriented towards the linking of interdisciplinary knowledge in many levels. Currently Ana Sabogal is the Director of Master Studies in Environmental Development and principal professor at the Pontificia Universidad Católica del Perú, at the Section of Geography and Environment.

List of Figures

List of Tables

1 Introduction: Definition of City and Public Spaces

Abstract

Nowadays most of the world population lives in the cities. This has been the case since the end of the twentieth century. Megacities are usual in countries with poor services and infrastructure which do not reach their rural areas. In this context, green areas fulfil a vital role, protecting the city's air, soil and water and allowing inhabitants to lead a healthy life. However, only in a few cases is this goal achieved, so it is becoming more necessary not only from the point of view of how urban ecosystems function but also considering inhabitants' need of leisure and health. To attain this, there is a search for city models in which emissions are calculated and reduced and ecosystems are restored and re-naturalized. The proposal implies re-thinking the hectic, fast-pace city life. Development of sustainable and livable cities is under debate. This chapter defines and discusses the concept of public spaces and green areas. The problems of Lima city are presented as well as the difficulty and need for Lima to have enough spaces to meet this need. Given the new challenges posed by the growth of cities and the changes in the relationship between the urban and the rural, we must consider the UN's Sustainable Development Goal 11, focusing on sustainable cities: how to "make cities and human settlements inclusive, safe, resilient and sustainable."

Keywords

City · Sustainable city · Public space · Green area

Since the end of the twenty century most of the world population lives in the cities. The megacities situated in countries with poor services and infrastructure where the green areas have not a priority. As result, there are a lack of green areas. The air, soil and water from the city are contaminated. This bring with a unhealthy city and diseases from the city population. In this chapter, development of sustainable and livable cities is under debate. This chapter defines and discusses the problems of Lima city. Given the new challenges posed by the growth of cities and the changes in the relationship between the urban and the rural, we must consider the UN's Sustainable Development Goal 11, focusing on sustainable cities: how to "make cities and human settlements inclusive, safe, resilient and sustainable."

1.1 Definition of the City

Since the end of the twenty century most of the world population lives in the cities. The megacities situated in countries with poor services

A. Sabogal, *Urban Ecology*, Sustainable Development Goals Series,
https://doi.org/10.1007/978-3-030-69905-5_1

and infrastructure where the green areas have not a priority. Even though city growth has occurred since the beginning of the industrial age, together with the development of industries, it is only since de 1950s that the population has moved to the city. 28.8% of the world population lived in the city in the 1950s. However, 46,4% were urban by the end of the twentieth century and in 2007 this figure reached 50% of the population (Endlicher 2012). While in the 1950s only 5 cities had more than 5 million inhabitants, by the beginning of the twenty-first century 16 cities had more than 10 million inhabitants (Kraas 2003 cit. Endlicher 2012: 245).

An important milestone in the development of the concept of city is the Athens Charter published in 1943 which set four guidelines for cities under the keynotes: living, working, recreation and circulation and suggested that green areas should be a functional part of the design. But it was not until the 1970s with the charter of the Club of Rome published in 1972 that there is a warning against the indiscriminate growth of the cities. Even though the world's most populated city is Tokyo with 35 million (2007) and an area of 13,551 km^2 (Hagan 2015: 125).

A mega city is one with a population of over 10 million inhabitants (Mertins 1992 cit. Endlicher 2012: 245). Several mega cities have developed in Latin America where population concentrates especially in capitals. Enormous metropolises such as Mexico City, Sao Pablo and Buenos Aires, small compared to other megacities, concentrate great part of the population in those countries. Lima is the 5th most populated city of Latin America (FAO 2015). This has occurred because they are the administrative centres of their countries. In addition to a high number of inhabitants, a megacity also shows a fragmented society, growing informality, among other problems (Kraas 2007 cit. Endlicher 2012: 247). In terms of population Lima is right at the limit not reaching yet 10 million inhabitants, but it is very difficult to manage due to the presence of great socioeconomic fragmentation and, therefore, can be considered a megacity. The classical development of a city is based on a grid. This form of development was inherited by a great number of cities in Latin America, such as Lima and Trujillo; walled cities which formed a grid pattern. Subsequently, with industrialization the city started developing around an urban centre and expanded in a concentric circle radiating from the centre.

The Spanish colonies in Latin America promptly established colonial cities typically traced with one- hundred-meter blocks like a chess board and one unbuilt block in the centre for the main square surrounded by the main government buildings (Heineberg 2017: 297). Properties of noble families were in the city centre and traditionally had an indoor patio. The distance from the centre depended on the economic power of the family; the workers and slaves were in the city periphery (Heineberg 2017: 298). This was exactly what Lima looked like when it was founded. Migration from the countryside to the city has played a significant role in its development. This migration has influenced both the city centre and the periphery. It has elevated the line of construction as well as caused the subdivision of blocks to accommodate new migrants (Heineberg 2017: 301). While the city grows the narrow streets widen often forming diagonal cuts (Heineberg 2017: 301). The city degrades in the periphery near the markets where low income population live (Heineberg 2017: 301). Houses form the middleclass are divided and small spaces are rented out with poor infrastructure (Heineberg 2017: 301). This becoming the so called "*Slums*" (Bähr and Mertins 1981 cit. Heineberg 2017: 303).

Little by little, Lima, once surrounded by defensive walls, has transformed itself and incorporated its surrounding areas. Thus, in the late nineteenth century the Parque de la Exposición was built outside the city walls, on what had been the Matamandinga Estate (Pacheco 2016: 13). The 42-hectare park, complete with marble columns, opened in 1872 with a grand exhibition of art and industry encompassing an array of themes. The park's architecture is part of the exhibition; among many impressive buildings, the Bizantine Pavillion stands out (Fig. 4.5). The park also featured a zoo stocked with wild and domestic species of plants and animals (Pacheco

2016: 31). The Las Palmeras Bandstand staged musical activities, such as orchestras and, later, phonograph performances; in the theatre there were and still are puppet shows; and a boating pond and cycling facilities provided space for sports activities (Pacheco 2016: 52). At the time of opening, admission to the park was not free; visitors had to pay a fee. This finally changed in 1921, under President Leguía, when the charge was scrapped and the park became a true public space maintained by the state (Pacheco 2016: 69). Throughout its history, the park has undergone dramatic events and transformations. In 1881, during the war with Chile, it was looted; sculptures were lost, trees felled (Pacheco 2016: 47). In 1898, the park was divided in two for the construction of roads for motorcars (Pacheco 2016: 53). Finally, the zoo was transferred in 1944 to the neighbourhood of Barranco, the old resort for the moneyed classes in the outskirts of Lima (Pacheco 2016: 41). At present, it falls within the city's boundaries. Thus, over little more than a century, the park has changed along with the city. The Parque de la Exposición has always been a centre of cultural and social activity. Today, it houses the Lima Museum of Art, and the theatre and pond are still there (Fig. 4.6).

The centre of Lima retains the grid structure introduced by Pizarro, as well as the structures of the colonial houses, though many have fallen into a state of disrepair. Lima possesses 608 historical monuments from the colonial era, all of which are worthy of preservation. It was for this reason that the historic center of Lima was declared a UNESCO World Heritage Site in 1998 (ONU Habitat 2015: 46). Many of the dwellings in the center are courtyard houses, characterized by passages that lead to the enclosed patios (Ledgard 2016: 245). The commercial life of the owners unfolded in these passageways, making them a transitional space between the public and the private (Ledgard 2016: 245). A major player in this part of the city is the hawker, who takes ownership of the streets (Ledgard 2016: 106). The use of the plazas, above all those in the center of Lima, as collective spaces for commercial exchange is rooted in traditions linked to the establishment of the migrant population. However, these practices are poorly managed, leading to a disharmony between population and planning (Ledgard 2016: 150).

The streets of Lima have become increasingly narrow over time. The avenues of old were between four and five meters wide, accommodating trees, benches and newspaper kiosks and providing space for conversation and an ice cream, whereas newer streets are just one and a half to two meters wide and devoid of trees. Thus, sidewalks have lost their role in recreation and socialization, being reduced to spaces for travel (Ludeña 2013:160). But the tight sidewalks are still used for the informal retail of a range of products; they host established businesses reselling the likes of used books or stolen goods, and ad hoc enterprises where one can find low-value items (Ludeña 2013: 149) such as socks, chocolates and many things besides. All this is part of the city.

In the 1950s the concept of city developed by urbanist Le Corbusier played an essential role in making the city more sustainable. Le Corbusier develops the concept of vertical gardens and green terraces as part of the architecture of housing complexes (Endlicher 2012). He outlines a completely different development from the model followed by the cities in the United States in the years shopping centres, industries and green spaces were built in the city outskirts forcing a dependency on a highway network (Endlicher 2012: 32). Nowadays, cities need to be rethought and new solutions found especially for megacities. The current debate about the impact of urbanization and the choice to decrease this impact lead to cities that give priority to densification of urbanization in order to make cities more compact (Catalan et al. 2008 cit. Francis and Chadwick 2013: 153). Lima is not exempt from this, even if it emerged as a city based on a model of detached single-family houses with front and back gardens—an ideal of the more affluent classes—resulting in a sprawling city of small houses (Ledgard 2016: 144). Lima has departed radically from these beginnings, having become an increasingly dense and gardenless city.

Following the trend of the time, new spaces for the working classes were planned in Lima. Until the 1930s, the *quinta* model predominated, with its origins in the spontaneous construction of alleys, without the involvement of architects (Ledgard 2016: 41). These "spontaneous alleys," in turn, stem from the huts on the estates where the rural laborers worked, in which services, such as running water, were shared. In later years there were attempts at constructing residential complexes, aimed at the working classes (Ledgard 2016: 41). It is in this context that open spaces and green areas emerged.

The Plan del Área Metropolitana de Lima y Callao, 2035 proposes the development of a compact city based on a denser built area (ONU Habitat 2015: 42). To this end, the aim is to initiate social housing projects that promote vertical growth (ONU Habitat 2015: 43). This plan proposes a polycentric city made up of 58 centralities (43). It also envisages the development of new poles of production, in Ancón and Lurín in Lima province, and the strengthening of existing production centers within the city, in Gamarra and Villa El Salvador (ONU Habitat 2015: 25).

Different city models have been proposed to end the dependency of the city on the countryside. A city uses great amount of energy resources and food producing a lot of waste and greenhouse gas emissions. It is impossible to continue in this way without causing further impact on climate change. Cities need to take responsibility for the pollution they produce. It has been suggested that the development of a city as an ecosystem would decrease consumption. This city would have certain independence and would be able to produce what it consumes, achieving a balance in terms of energy. Size would be an essential factor to reach this goal. So that a city can be considered ecological, it should respond to five planning principles: energy consumption decrease, waste reduction and recycling, protection of livelihood, conservation and promotion of nature (Sukopp and Wittig 1993 cit. Endlicher 2012: 178–179).

Cities are constantly transforming. Therefore, a city should be resilient to change. To this effect the Seoul Energy Declaration 2007 targets city resilience by proposing it can be attained through energetic terms (Kinz and Choi 2011 cit. Bulkeley et al. 2014). However, this is not enough to preserve the city ecosystems. It is essential to have small spaces with a diversity of ecosystems (Sukopp and Wittig 1993 cit. Endlicher 2012: 178–179).

Transformations of cities are dependent not only on technical specifications; their development and transformation are marked by the populations of which they are composed. In Peru as elsewhere, urban migration used to mean migration to the large metropolises, especially the capital. This was true of the influxes into Lima during the 1950s and the 1980s. At present, we are witnessing a change. To a large extent, present-day migration entails relocation to the so-called intermediate cities. This applies as much to Peru as it does to other countries. Thus, in Brazil, for example, the mid-1990s ushered in a second cycle of urbanization, which boosted both the number and the population of medium-sized cities (Angeoletto et al. 2019).

The typology of Lima must respond to the city's new uses and users (Ledgard 2016: 247). It is worth noting that the center of Lima hosts many of the city's economic and administrative activities, and thus a large share of its jobs (ONU Habitat 2015: 44). Thus, the centre must not become a museum of old houses that cannot be remodelled or used (Ledgard 2016: 246). To ensure its continuity, the city must incorporate the uses the population makes of it. The development of Lima has been accompanied by the densification of housing due to larger family sizes.

Urban ecology is the science that studies living systems in the city. Therefore, it includes all components of the ecosystem such as the water, the ground and the living beings constituting interrelated systems. It is important to point out that unlike natural ecosystems, plants and animals are introduced into urban ecosystems finding refuge where the trophic chain has fewer predators compared to natural areas. On the other hand, many domestic animals introduced and liberated into the city find a new space and renaturalize. New taxa and habitats are born as well as new city spaces which would not exist otherwise (Wittig 2002 cit. Endlicher 2012).

1.2 Sustainable City

The crisis in the cities generated by population growth makes us question what a sustainable city is. City resilience is undoubtably a key for their survival, but it is not the only factor. Great part of the problem lies in the reduction of waste which means a change in lifestyle; a slower lifestyle would give people time to produce their own food and recycle waste. Given the new challenges posed by the growth of cities and the changes in the relationship between the urban and the rural, we must consider the UN's Sustainable Development Goal 11, focusing on sustainable cities: how to "make cities and human settlements inclusive, safe, resilient and sustainable."

The current climate crisis is, to a great extent, a result of the inadequate management of the cities. Not only now but always have cities been national and international migration centres. The development of *gettos* and slums are part of the city dynamics; places that slowly integrate to the city but are linked to economic informality; informal activities which are not taxed but generate important processes that lead to emancipation (Endlicher 2012: 252). In megacities like Lima informality is seen in informal constructions; precarious occupation of land located in risky areas; occupation of marginal urban spaces and spaces allotted for green areas.

In Lima, as in many other cities and towns in Peru, the traditional urbanization process involves various stages. First, clay is removed from agricultural soil for the production of bricks. Then, after agricultural land loses its fertility and its value, it is divided into lots and sold unofficially at very low prices by so-called land traffickers. The sale gives rise to squatting, in which the buyers situate themselves on the site and erect shacks of matting to defend it. After a population has taken root, they request the installation of basic services and, little by little, each occupant develops their land based on their means; it is only once this has been done that plans are drawn up and submitted to the municipality. Constructions tends to remain incomplete for lengthy spells, which keeps tax rates low. While Lima was developed on flat land, migrant populations have gradually established squatter settlements on the hills that surround it. Though environmentally unstable and susceptible to landslides, these areas have become officially recognized neighborhoods connected to the rest of the city by steep staircases (Ludeña 2013: 169).

In this context of constant urban change and alteration, it is interesting to ponder the concept of spatial justice and to apply it to city development. This is difficult to achieve because the conception of and attachment to space differs from culture to culture, and tends to be more pronounced among rural populations (Lister 2013). Cities, especially younger ones of the kind found in Latin America, have large migrant populations characterized by their detachment to the urban space. To advance toward the sustainability of cities, we must first analyze the characteristics of the population and integrate their needs and attachments with the conception of space. The sustainability of a city thus depends on sustainable planning and construction, but also on achieving a connection with the inhabitants. If managed judiciously, this identification can present an enormous advantage in terms of planning and restructuring, taking us closer to spatial as well as environmental justice in the city.

Access to green areas is a topical aspect of environmental justice. In the specific case of Lima, public spaces are greater in number and better distributed in the middle-class neighborhoods (Sabogal et al. 2019). Moreover, biodiversity tends to be greater in affluent neighborhoods, where people are more likely to afford more exotic species of plants (Aleixo et al. 2016 cit. Angeoletto et al. 2019).

Since cities are constantly changing, it is interesting to consider the concept of spatial justice so that it can be applied during the development of the city. However, spatial justice is difficult to apply because the conceptualization and attachment to space is different in each culture and it is very strong among rural population (Lister 2013). Cities, especially the newly formed as is the case of the ones in Latin America, are inhabited by a great number of migrants who are unattached to city space. If we study the characteristics of the population and we manage to consider their needs and affections in relationship to the concept of space, we could move forward in terms of sustainability. City sustainability not only depends on sustainable planning and construction but also on making inhabitants feel identified with the city. This, if adequately managed, would mean a great advantage in terms of planning and restructuring. Furthermore, it would help the city reach spatial justice, which includes environmental justice.

Undoubtedly sociocultural aspects play an important role in the development of a city which is redesigned by its population day after day. The absence of the government in the cities in developing countries has made the population, especially in marginal areas, self-organize to modify and manage spaces and create new ones (Hardoy and Satterthwaite 1986 cit. Hagan 2015: 147). It is important to point out that in 2003 43% of the population in the third world lived in "the slums" whereas only 6% did so in the developed countries (UN 2003 cit. Hagan 2015: 147).

Cities can follow two models of organization; ones where the centre is important and everything revolves around it; and others with independent neighbourhoods where people do not have to commute for hours from their homes to their place of work, stuck in traffic every day. Most Latin American cities follow the latter form of organization; centralized cities revolving around a point, where low income people live in the periphery and the wealthier live in secluded large spaces. In these cities, the centre is both an administrative and a commercial centre.

Considering city density, there are two city models; the garden city, where there are interconnected green spaces and ecological corridors as well as great distance between constructions to allow for plenty of green spaces; the compact city, with a lot of buildings and an interconnection with other cities forming microcities (Hagan 2015). The decision regarding the type of city adequate for the space depends on several factors such as ecosystem, water availability, land price and the sociocultural characteristics of its inhabitants. Both models produce very different results in terms of density. For example, whereas density in London is 330 inhabitants/ha., Hong Kong's is 5000 (Hagan 2015: 49). The former is a Garden City and the latter a Compact City.

For city models to be sustainable they need to consider natural resources and ecosystems. Most Latin American cities have a highly dense centre which decreases in the periphery giving them the name of "centre-peripheric". Cities like Madrid and many European cities in the Ruhr basin have many urban nucleuses provided with complete services, decreasing the pressure on the city centre and reducing inequality between the city and its periphery. However, in desertic ecosystems like Lima it is difficult to think of and expanded city due to lack of water and the high cost of implementing networks and services.

The ideal city proposed by Le Corbusier is a compact city surrounded by a food producing belt which could be used as a buffer for the absorption of city waste (Hagan 2015: 47). In this high-density city Le Corbusier would design buildings spread across green areas, escaping from the relationship between density and city quality (DETR 1997 cit. Hagan 2015: 48). The debate about sustainable density depends on the view of the age. In the twentieth century, after Le Corbusier, there is a proposal that the city should be vertical and dense with intertwined green areas (Hagan 2015: 111). Nowadays this translates in the ratio between square meters of green areas per inhabitants. Under this vision, a sustainable city is a high density, poli-centralized city with coordinated functionality and little dependency on the car (Rogers 1995 cit. Endlicher 2012: 179).

As noted earlier, the Plan del Área Metropolitana de Lima y Callao, 2035 proposes the creation of a polycentric city made up of 58 centralities (ONU Habitat 2015: 43). It also envisages the development of new poles of production, in Ancón and Lurín, and the strengthening of existing production centres in Gamarra and Villa El Salvador (ONU Habitat 2015: 25). If well managed this plan could help to reduce heavy traffic within the city, provided that it is accompanied by the development of housing and spaces for the population. Conversely, inadequate governance could result in even greater class polarization than is already in evidence in Lima.

The city development models as of the twentieth century concentrate on the balance of emissions, which means that the city aims to have a balance between its emissions and their absorption by planting trees, thus, achieving urban metabolism. This is linked to people's lifestyle by trying to shorten daily distances and enabling a city where people can walk and not rely on their cars (Hagan 2015: 54). In this sense, city density reduces the Ecological footprint (Rees and Wackernagel 1996 cit. Newman and Waldron 2013: 108). Whereas the city models that seek to increase the area of each house in the periphery lead to an increase in the use of cars and consequently an increase in the city's Ecological Footprint (Newman and Waldron 2013: 114). Urban metabolism aspires toward a balance between emissions and the consequent use of fossil fuels, dependence on external materials, and enormous production of waste (Angeoletto et al. 2019). Therefore, according to Odum's classic definition, cities are heterotrophic ecosystems, dependent on external energy (Odum and Warret 2006). When it comes to Lima, this energy depends on agricultural and mining products, which enter the city via the Carretera Central highway. Thus, in cases of landslides, rivers bursting their banks, or highway blockades, food prices rise exorbitantly. This is one of the city's points of vulnerability.

The transformation of peri-urban areas continues as Lima grows. The Oquendo Estate, located very close to the international airport in Callao, forms the northern periphery of the city of Lima. Situated near the main thoroughfare used to transport produce from northern Peru to the capital, in an area suited to vegetable production, little by little the estate was divested of its cropland. Following the agrarian reforms of 1969, ownership passed to the workers, who gradually sold off the land. Up to the end of the 1990s the estate still contained some agricultural land (as well as the old estate house), having received support from various international initiatives for its conservation and crop production, but by now it has been entirely divided into lots and precariously urbanized—despite still lacking paved roads. The estate, as it is still known locally, is now controlled by gangs, and is blighted by both crime and poverty. Other areas on the margins of the city have also been divided into lots, losing their agricultural usage; this has denied inhabitants the income enabled by adequate land management, and Lima a supply of vegetables that would contribute to sustainable urban metabolism.

Two types of models are suggested to reach a sustainable city; on one hand, a decentralized model with multiple urban centres. On the other hand, cities with an ample transportation system. The city should encourage inhabitants to walk which is directly related to their health (Newman and Waldron 2013: 114). Cities that encourage walking should meet three requirements: density, connectivity and proximity (Frank and Engelke 2005 cit. Newman and Waldron 2013: 117). Pedestrian streets with no cars are common in Europe (Newman and Waldron 2013: 116). There, people walk three times more than in North American (Basset et al. 2008 cit. Newman and Waldron 2013: 114). The possibility to walk allows more interaction among citizens and contributes to spatial equity (Newman and Waldron 2013: 121). In big metropolises, especially in South American cities such as Lima, lack of security and unfriendly sidewalks for pedestrians or bicycles make it difficult to implement spatial justice.

The growth dynamic in a city has cycles. They are space temporal dynamics that start with urban expansion from the centre, followed by a period of coalition that takes place when an area is saturated with constructions. While the city expands it includes incrusted green areas and unites with different areas that were previously disperse; finally, the same cycle can repeat itself in another area (Xu et al. 2007 cit. Francis and Chadwick 2013: 32). Lima has grown from the centre to the periphery; it started to grow in Rimac river valley and expanded to the beach areas of Miraflores and Barranco. Currently it has subsumed Chillon river valley and it expands to Lurin river valley getting water supply from these three rivers. However, there are areas which cannot expand evenly because of their physical nature as explained by Xu et al. cit. Francis and Chadwick (2013: 34). In Lima there are small hills like San Cristóbal, Cantagallo or hills in La Molina or at the edge of the Rimac river. These areas, where risks concentrate, are usually temporarily occupied by low income migrant population. Normally, the city densifies with constructions from the centre displacing green areas. However, these geographical barriers halter the process in certain spaces leaving some natural areas (Jim 2010 cit. Francis and Chadwick 2013: 36). Undoubtedly, this can be modified by politics and urban planning in correlation with taxes and space restrictions. Political decisions have to take into consideration environmental aspects such as wind direction, valley structure or river banks, and especially they have to define and manage the risk zones in order to avoid possible environmental risks such as river banks and pollutant concentrations on the slopes of the hills.

The growth dynamic of a city leads to the development of marginal neighbourhoods. In 2010, about 828 million people lived in different marginal neighbourhoods. On one hand, migrants place themselves outside the city where the infrastructure is not developed and the green areas are invaded and on the other hand, a great number of people live in crowded, overpopulated areas where services are degraded and insufficient. Both spaces are culturally different.

The periphery of a city often houses wealthy population looking for more space and green areas as well as low income migrants from the countryside who invade the areas. This contrast makes planning in the periphery difficult. The periphery can become dependent on the city or develop its own dynamic, contributing with green areas, parks and crops and supplying the city (Hagan 2015: 71). Suburban areas can be self-sufficient developing services, parks, markets and shops (Hagan 2015: 75). The area could have a farm strip for crops to supply the city just as proposed by Le Corbusier.

When reflecting about the city we question now more than ever how an ideal city develops. A city that comprises ecology, urban metabolism and does not pollute, with sociocultural spaces where the population can develop and live ideally. An ideal city is one which we do not have to think about constantly. However, this is not enough. We should like to live in it, we should feel comfortable and we should know and relate to nature. Lima, for example, could be a beautiful city that integrates ocean, river and history, where ecological corridors enable us to walk and contemplate nature without having to escape the city (see Fig. 1.1).

An interesting attempt to develop a sustainable space for the city of Lima was the Huaycan project; launched by the municipality of Lima in 1984, this community and self-management initiative is situated to the east of the city, by the seventeen-kilometre stage of the Carretera Central (Ledgard 2016: 118). Water was drawn from the subsoil of the Rimac basin, and wells were planned for the lower section to harness energy from the upper part. The local population took charge of making the building materials (Ledgard 2016: 133). This model was based on traditional Andean communal organization, with housing units surrounding a central plaza and designs discussed and approved by a community assembly (Ledgard 2016: 120). The inhabitants cleaned and prepared the site, decided on and applied the design, and self-managed the project (Ledgard 2016: 124). Thus, the design has evolved over time (Ledgard 2016: 125). Although this particular project was overseen by the municipality, it reflects a self-management of space that characterizes a large proportion of Lima's marginalized areas. This self-management extends to parks and gardens with fruit trees and other edible species (Sabogal and Martínez 2015). As in the case of Huaycan, this is often combined with the husbandry of small animals, such as guinea pigs, for sale or local

Fig. 1.1 Rimac cliff. *Author* Ana Sabogal

consumption. But the large, unforeseen risk in the Huaycan project was its proximity to the stream, given its propensity to increase in volume during the rainy season. And although the project involved the population in self-management, development stagnated and today the neighbourhood is little different than Lima's other *pueblos jóvenes,* or shanty towns.

1.3 Public Space

A Public Space is a place of congregation and entertainment with or without green areas. Public spaces include natural spaces together with squares and sports areas. If we consider that Public Spaces are places of interaction, the city is a Public Space with parks, squares, cafés among others. This concept goes in hand with walking and using the city beyond its private spaces (Newman and Waldron 2013).

Public spaces are open spaces with free passage. They contribute to interculturality, tolerance and acceptance of social and cultural diversity in the city. There are spaces for social cultural encounters where multiple cultures and social classes meet. In this sense, they have an important function to integrate the population in the city and resolve social problems. Public Spaces are not only a reflection of society, but they can also transform it. Public spaces are also political spaces where social problems emerge and the solutions of them are find. They are often the place where social struggles and transformation occur.

Another interesting case is Lima's Chinatown (known locally as *Barrio Chino*). This neighbourhood is much visited by locals of Chinese and non-Chinese origin, and constitutes a genuine public space from a social and cultural point of view where *Limeños* of different classes converge. Chinese immigrants, mostly from Canton, began arriving in Peru in the mid-nineteenth century, as a source of rural labour after the emancipation of African slaves. Today Peru's population of Chinese descent is very large, amounting to some three million people—around ten percent of the total population (Chuhue 2016: 11). Chinatown is located in the historical centre of Lima, within the area of Barrios Altos. Found there is the Chinese temple and oracle of I-Ching, built in 1868, where many *Limeños* go to have their fortunes told, as well as Chinese restaurants, herbalists, retailers and, until relatively recently, opium dens and a school for the Chinese community (Chuhue 2016). It is a space where different cultures and social classes—from the Andean to the Chinese—come together and intermingle.

There are different types of public spaces based on size. Among them we can find:

- Squares: They are open spaces often located at street intersections, usually round covered with pavement, tile or cement with few or no green spaces. They are focal points in the city mainly used for public events. Squares are very often found in Spain and Latin America.
- Park: Open spaces with free access to the population. They have green areas that can be used by neighbours or municipalities as public spaces. They can be small or big and based on their size, they are considered municipal or zonal. They have multiple functions. However, the most important ones are providing entertainment and cleaning the air in the city.
- Sports Areas: They are public or private spaces with sports facilities. Green areas are secondary. Many sport areas are clubs or associations.

When we talk about public spaces, we do not mean only sidewalks or street shoulders, but all the places where citizens deliberately spend their free time. Citizens can interact with others or simply stroll along to rest from the hectic city life. These places are where citizens put aside their private sphere.

Returning to our description of Lima, one can identify:

- Plazas: In Peru there is a long tradition, predating the Spanish conquest and rooted in Andean customs, of using public spaces for commercial exchange. In the Andes, people from rural villages and areas descend upon agricultural markets and shows at town plazas to buy, sell or exchange their produce. It is here too that social activities, such as parties and gatherings, are held. Disposed around the plaza are many of the local jurisdiction's main public buildings—such as the municipal building, the seat of parliament, and the cathedral in the case of Lima's Plaza de Armas. The tradition of using plazas as spaces for commerce and recreation travelled with the migrant populations to Lima, where the plazas have an importance that transcends official events. Indeed, the environs of the Plaza de Armas hosts dynamic informal commerce that forms part of the life of the city. Here, social tensions pose challenges that go beyond the formal channels of the city as a whole. Added to this is the series of artistic activities that unfold on the plazas, such as juggling, street theatre, and busking, among many others. All are forms of social organization that characterize Lima and are accepted by society.
- Parks: Lima has two levels of park management. First, municipal parks are smaller spaces used by adults and children for their everyday leisure needs. Many of these parks have children's play equipment and basic sports facilities, where residents can interact and engage in activities with each other. These parks are relatively safe spaces, and are often monitored by private or municipal security guards. Second, the zonal parks often span several municipalities, and are administered by the Municipality of Lima through the Parks and Gardens Service. These parks are on a larger scale and enable greater interaction.
- Sports facilities: Lima's sports facilities are few in number and organized along class lines; one exception is the soccer stadiums, where different social classes intermix, but this is a space for spectating rather than participating. Most sports facilities are confined to clubs catering for higher social classes; examples include the Club de Regatas de Lima, a rowing club; and the Club Germania, which specializes in swimming. Thus, social classes interact with their own, without coming into direct contact with cultural or socioeconomic distinctions. Soccer is a passion of the less well-off classes; many children play in the hope of imitating the world-famous superstars of the game. The zonal parks, described in Chap. 4, play an important role as sports grounds—most notably for soccer.

Here, children interact and enjoy their escape from family and societal pressures. Sports grounds, essential in the *pueblos jóvenes*, play a very important role in the socialization of young people by way of soccer, local fairs, fiestas, and other events.

1.4 Green Areas

They are public spaces where a high percentage is covered by plants. While parks are big, gardens are smaller and can be public or private. Green areas have two purposes: air purification and entertainment for the city population.

Green areas have a very important social role which is to close social and cultural differences and free the mind of people towards art. They contribute with physical and mental health of inhabitants as well as their social integration. In them some people do sports, others stroll with their children or just read. It is a place to find peace and take refuge from family problems or just reflect without talking to anyone. We change space and attitude to one of contemplation of the landscape, flowers and birds leaving behind daily chores and allowing reflection.

The idea that plants are needed in the city became imminent when the cities began growing uncontrollably and became overcrowded. The usefulness of the parks to preserve the health of working people through sports and entertainment arises because of the industrial revolution which triggers the need of the parks as public space.

In Lima, as a trend that followed the Athens Charter, residential complexes were designed and built around green spaces. The first such space in Lima was aimed at the city's working classes. Residential Complex Number 3 was built in 1946 in one of Lima's industrial areas, in central Lima. Conceived in response to the debate about minimum areas per family and green spaces, it contains a central green area with communal services and pedestrian access bordered by gardens (Ledgard 2016: 44). Other representative examples following the same trend, but constructed many years later, were aimed more at the middle classes. They include the Santa Cruz Residential Complex, opened in 1966; and the San Felipe Complex, in 1968. Both designs feature green spaces bordering buildings and common areas. Today, the expansion of Lima and the construction boom has meant new buildings are appearing again; however, these are mainly small apartment blocks without green spaces or scope for communal or social urban life, where young people spend their free time alone and not in the company of neighbours or in the neighbourhood.

In "*Plano Voisin*" designed by Le Corbusier for the city of Paris in 1925, which led to the Athens charter, green area is used as background for the design; this proposal modifies the concept of city totally. According to it, free spaces should occupy more than 80% of the design (Vercelloni and Vercelloni 2010: 242). This idea of the park as a public space has prevailed to this day. The landscapist Ludwig Lesser (1869–1957) developed and implemented the concept of Green Area as public spaces and defines the People's Park *(Volkspark)* as one that has large spaces suitable for games and sports with grass and shade giving trees, where citizens of any social class can go, meet to find peace and quiet and get away from city daily stress (Vercelloni and Vercelloni 2010: 20). This idea was implemented in the Ludwig Lesser park in Berlin, under the concept of *Volkspark*, known today as Zonal Parks, big parks with a lot of space for sports, games for children, among others (see Fig. 1.2).

In these spaces the population can find even more freedom that in the closed spaces of the modern city where there are not enough square meters per inhabitant or where living with others poses certain restrictions. These spaces also help decrease social problems in the city since the citizens need to abide by certain social rules. The use of these spaces depends on each culture; whereas Turkish and Moroccan migrants in the Netherlands use green areas for family group orchards, local population uses them individually for walking or cycling (Buijs et al. 2009 cit. Francis and Chadwick 2013: 159). However, if badly managed and when there is no consideration of the population's social role, parks can

Fig. 1.2 Ludwig Lesser Park, Berlin. *Author* Ana Sabogal

become spaces where social problems are released. They could even become unsafe places with vandalism. A situation like this can lead to an increase of closed spaces transforming parks in places with fences which are opened only for a few hours a day or they may even be privatized by the neighbours. This often happens in Latin American cities.

Lima's peripheral areas, such as those alongside the River Rimac or the sea, present great potential for urban development, but at present their populations are largely marginalized. These spaces often function as retreats for those on the fringes, such as the homeless, drug addicts, and alcoholics (Ludeña 2013: 164). Meanwhile, parks in residential areas are increasingly becoming spaces enclosed by railings, gates or other barriers intended to exclude those not from that neighborhood. Lima's more affluent residential districts, such as La Molina, are characterized by a lack of sidewalks (most residents own cars), gates, and private security to ensure that non-residents stay away.

Leberecht Migge (1881–1935) in his book "Garden Culture of the Twentieth Century" published in 1931, develops concepts for urban parks which rule nowadays. He introduces the idea that parks are useful and should offer entertainment and health for citizens. Parks should have sports facilities, paths for walking, nice views for entertainment enhancing the geometry of the design (Kluckert 2000: 476). Green areas also clean the air and moderate climate. The idea that green areas are important to protect the good health of the citizens is also developed by Martín Wagner, in his book "*Das sanitäre Grün der Städte*" (The Sanitary Function of Green in the Cities), published in 1915 who postulated that parks purify the air in the city (Kluckert 2000: 477).

But the notion of the park as a public space remains alive in Lima's zonal parks. At present the city has eleven such parks, encompassing a total of 371.80 hectares as of 2010 (Ludeña 2013: 98). Lima's first zonal park was Cahuide, a 21-hectare site that opened in 1971. Located in an underprivileged part of Lima that otherwise lacks green areas, the park with its multiple spaces and sports facilities plays an important role in environmental health and justice.

The history of gardens later called green areas is the history of culture. Gardens are born in cities where man is away from nature. Garden design corresponds to a perception of the world. Each culture develops their own vision with certain qualities. Many times, religion and man's quest for god when facing life's challenges is represented in gardens. Many of the first gardens were cemeteries. Now, in the twenty-first century it is still perceived that the design of parks and green areas in the city is linked to holism and seeks to incorporate art.

With the loss of proximity to nature and as the world becomes more globalized there is a return to what is indigenous of an area in search for an identity with nature. This search for what is natural has influenced the artistic design of the parks and gardens since the late nineteenth century. However, it is not until the end of the twentieth century that the sustainability of the green areas and gardens in the city is proposed and this goes in hand with considering climate and plant selection. From then on, ecological principles are applied in the design of parks considering not only plants but in general the development of a sustainable ecosystem. A pioneer in this new type of garden was without a doubt Roberto Burle Marx (1909–1994), with his masterpiece "Parque del Este" in Caracas, where he includes xerophyte, aquatic and forest gardens also applying the studies of nature captured in the Humboldt Planetarium. Once more, the work of Burle Marx incorporates the ecological principles in the Flamenco Park in Rio de Janeiro where the beach and even some houses are integrated to the design, showing an interesting combination between the ecological and social aspect throughout 122 hectares.

Another characteristic of the twentieth century gardens is the incorporation of the surroundings to its design as a legacy of the landscape gardens. Landscapers study the landscape from different perspectives, focusing either on biological or on architectonical aspects. The garden of Saint Clotilde in Barcelona enhances the beauty of its surroundings and incorporates the ocean and the beaches to its design (Kluckert 2000: 471). The incorporation of the surroundings to the landscape was seen before in the work of Peter Joseph Lenné, who designed gardens around beautiful sights (see Chap. 3). On the other hand, Patrick Geddes (1854–1932) tries to perceive and rediscover the landscape's geological aspects and combine design with social aspects; he mixes different methods of geography, biology, agronomy, geology, history and culture of the place, developing an interdisciplinary method (Vercelloni and Vercelloni 2010). He intends to merge geology which is physical to what is historical, combining theatre, greenhouses, fountains, a cathedral, designs in which history and modernity are intertwined. Finally, the proposal made by Isamu Noguchi (1904–1988) who introduces sculpting of natural elements such as marble and stone incorporating geology, geography and history to garden design. Thus, he suggests that nature can created art (Vercelloni and Vercelloni 2010: 245).

The Villette Park, designed by Tschumi, transfers the focal point of design from the architectonic to the surroundings (Vercelloni and Vercelloni 2010: 248), followed by the restoration of abandoned spaces in the city: train rails, abandoned factories are turned into parks returning to a poetic vision. When doing so, Gilles Clément develops the theory of "the garden in movement" considering nature is in constant change. He conceives the idea of taking advantage of energy in space and promotes respect to plants that grow without being planted, in the wild calling for their rightful space in the park (Vercelloni and Vercelloni 2010: 251–253). A clear example is André Citroën park in Paris, built over the old car factory by Patriel Berger, Gilles Clément, Viquier Jordi and Alian Provost; this park picks up the history of the place when it recreates the watering channels and greenhouses for cultivation (Vercelloni and Vercelloni 2010: 249). What cannot be overlooked about Lima is its 300 *huacas* (ONU Habitat 2015: 46), some of them incorporated into the design of the city as cultural spaces.

Since the twentieth century the concept of "*Urban parks*" proposed by de Jean-Claude Nicolas Forestier (1861–1930), forest engineer, combines formal design and nature, returning to the idea of gardens as collective space. He defended and proposed the restauration of forests outside Paris, Bois de Vincennes and Bois de Boulogne, as an important part of the city of Paris (Vercelloni and Vercelloni 2010). The redesign and defence of these forests as well as the conception of a park system for Paris are an important part of his work. He did not only defend the existing forests, but he designed a bike lane for cyclists in the Vincennes forests and he began the discussion about ecological corridors. Frank Lloyd Wright conceives a green city, where nature surrounds architecture immerse in it, green area is seen as an indistinct background of design and seen from the window as a spectacle in which human beings are not taking place (Vercelloni and Vercelloni 2010: 240), integrating surroundings and design so that they are not separate spaces.

Lima's Pantanos de Villa Wildlife Park is a 276-hectare site within the city boundaries, and one of its few ecological spaces for the conservation of flora and fauna. Hover, this enclosed, protected space is gradually being diminished by construction. And although the park's conservational role is undeniable, it is not fully open to the public—access is restricted to schools and universities for environmental education—without harmonization between both needs. There have been some recent attempts at improving access; boating areas have been introduced, for instance, but this renders the fauna a spectacle to be seen from afar, in the interests of conservation, and the space is still far from being accessible.

Calculation of green areas includes both public and private spaces, such as green sidewalks, street shoulders, river ledges and even the ocean. Thus, green spaces can have a diversity of characteristics. They could range from absolute urban areas to peripheric areas where ecological systems and social composition are completely different and, consequently, so is the purpose of public spaces and green areas. As mentioned before, green areas are a public space. But in Peru there is no legal definition for green area; when we speak about public spaces, there is no distinction between spaces with vegetation and or spaces without; moreover, terms are often used as synonyms. However, the quality of the city is measured based on the number of square meters of green areas per inhabitant. Whereas, for health reasons, the WHO recommends at least one park or public space every half a kilometre so that the inhabitants can reach the parks on foot (United Nation 2012). This concept implies the need for connections between spaces. If ecological corridors or interconnections were added between parks, it would enable pedestrians to walk in the city and the number of motor vehicles would decrease as well as the quantity of gas emissions, improving both the air quality and health of inhabitants.

In all cities, green and public spaces and their usage depends on the local culture. In Lima, where a large section of the Andean migrant population is habituated to life in elevated areas, the stairways ascending the inhabited hills have become interesting features of the public space. Thus, in 2010 Lima had 91.63 hectares of stairways, compared with 59.59 hectares of sports grounds (Ludeña 2013: 98). Although strictly speaking the stairways do not play a role in the city's environmental health, if well designed and tree-lined, they can contribute to it.

As mentioned in the paragraph above, green areas are important for environmental health. The amount of green space varies from country to country. Moreover, it is difficult to make a comparison between continents because the form of measurement is not the same in every country. Some countries calculate the amount of green areas as a percentage of the city's surface, whereas, others, like Lima, do the calculation as a percentage per inhabitant. Finally, some calculate the average distance that a citizen needs to travel to reach a green area. Whereas, calculating square meters per inhabitant is generic, considering the city's total percentage of green areas or the distance to green areas reveals the city's quality.

As mentioned before, the amount of green area in each city varies a lot. For example, Lima has fewer than 3 m^2 of green areas per inhabitant, while Hong Kong had 3 m^2 of green area per inhabitant by the year 2004 and Singapur 7 m^2/inhabitant which is more than double (Jim and Chen 2008 cit. Francis and Chadwick 2013: 73–74). In Berlin the percentage is around 34,8%, and 150 m^2/inhabitant. In Nueva York it is 26.8% and 25 m^2/inhabitant. In Mexico DF it is 8,9% and 15 m^2/inhabitant (Wessolek 2010 cit. Endlicher 2012: 108). In Great Britain, 14% of urban areas are green spaces (Sadler et al. 2010 cit. Francis and Chadwick 2013).

There is a relationship between the amount of tree vegetation in green spaces and socio-economic level of the area (Young and Jarvis 2001 cit. Francis and Chadwick 2013: 48). However, in Lima the middle class is the group with the highest percentage of public green areas (Sabogal et al. 2019). This has an influence on the number of birds (Fernández-Jucic 2004 cit.

Francis and Chadwick 2013:48) which help disperse seeds and are important for the entertainment of city dwellers.

As noted earlier, among Lima's middle-class areas the model of detached single-family houses with front and back yards (Ledgard 2016): reigns, resulting in a sprawling city of small houses (Ledgard 2016: 144). And while many front yards have survived, most today are closed off. Lima's front yards were once the elements of a city of private green spaces, designed and maintained by inhabitants. But this dynamic has shifted dramatically toward the uncontrolled construction of corporate buildings, increasing density and replacing residential yards with uniform, styleless gardens. However, it is still interesting to note the calculations for the city's green areas.

In Peru, specifically in Lima, the high cost of maintenance together with the low level of tax collection generate a correlation between the amount of green areas, both public and private, and the socioeconomic level. District with high incomes have more green areas like San Isidro with 22,09 m^2/inhabitant., San Borja, a middle-class district, has 11,86 m^2/inhabitant. Miraflores, a mixed area, has 13,84 m^2/inhabitant whereas low income districts such as Villa María del Triunfo have 0,37 m^2/inhabitant, Breña 0,37 m^2/inhabitant or San Juan de Miraflores 1,65 m^2/inhabitant (INEI 2018 cit. SINIA 2020).

Although these are official statistics they are not a true reflection of reality, as many areas set aside for parks were ultimately lost to squatters or put to different uses. Moreover, Lima's lack of water means that many parks are not green spaces at all, having been left to dry out or even paved over (Ludeña 2013: 175).

References

Angeoletto F, Fellowes M, Essi L, Santos J, Johann J, da Silva D, Moraes N (2019) Ecología urbana y planificación: una convergencia ineludible Austin, G (2014) Green infrastructure for Landscape Planning. Integrating human and natural systems. Rutger, Glasgow, p 266

Bulkeley H, Castan Broto V, Edwards G (2014) Towards Low Carbon Urbanism "from Local Environment" (2012). In: Wheeler S, Beatley T (eds) (2014) The sustainable urban development reader, 3rd edn. Routledge, London, New York, pp 101–106

Chuhue, R (2016) Capón: el Barrio Chino de Lima. Munilibros 2. Municipalidad de Lima, p 87

Endlicher W, Kraas F, Krajewski Ch (2012) Einführung in die Stadökologie. UTB, Stuttgart, p 272

FAO (2015) Agricultura urbana y periurbana en América Latina y el Caribe. https://www.fao.org/ag/agp/greenercities/es/CMVALC/lima.html. Accessed 24 July 2020

Francis R, Chadwick M (2013) Urban ecosystems: understanding the Human Environment. Routledge, USA, p 220

Hagan S (2015) Ecological Urbanism: the nature of the city. Routledge, Oxon, p 174

Heineberg H (2017) Stadtsgeographie, 5th edn. Utb, Germany, p 504

Kluckert E (2000) Grandes jardines de Europa: desde la antigüedad hasta nuestros días. Könneman, Colonia, p 496

Ledgard R (2016) La ciudad moderna: textos sobre arquitectura peruana, 2nd edn. PUCP, p 285

Lister N-M (2013) Map-maping as place-making: building social capital for urban. In: Dale A, Dushenko W, Robinson P (eds) Urban sustainability: reconnecting space. University of Toronto Press. Toronto, Buffalo, London, pp 30–80

Ludeña W (2013) Lima y Espacios Públicos, perfiles y estadística integrada 2010. PUCP, p 224

Newman L, Waldron L (2013) Towards walkable urban neighbourhoods. In: Dale A, Dushenko W, Robinson P (eds) Urban sustainability: reconnecting space. University of Toronto Press. Toronto, Buffalo, London, pp 106–126

Odum E, Warret G (2006) Fundamentos de Ecología, p 598

ONU Habitat (2015) PLAM 2035, sistematización del Plan del Área Metropolitana de Lima y Callao, p 2035

Pacheco J (2016) Parque de la Exposición: el jardín de Lima. Munilibros 5. Municipalidad de Lima, p 90

Sabogal A, Martínez M (2015) A study of ecological corridors in two quarters of lima: Chorrillos and Villa El Salvador. Perspect Global Dev Technol 14 (2015):587–596

Sabogal A, Cuentas MA, Tavera T (2019) Espacios públicos: Estudio en el distrito de Santiago de Surco en Lima, Perú. Revista Kawsaypacha 3:105–138

SINIA (Sistema Nacional de Información Ambiental) (2020) https://sinia.minam.gob.pe/indicador/998 Revised: 14.02.2020

UNEP (2012) United Nation Environmental Program. Cities and Carbon Finance: a feasibility study on an urban CDM, p 84

Vercelloni M, Vercelloni V (2010) Geschichte der GArtenkultur von der Antike bis heute. WBG, Darmstad, p 275

2 Landscaping Study and Methodology

Abstract

It is necessary to know the principles of landscaping in order to design green areas in the city and make Lima a more inclusive and resilient city according to the Sustainable Development Goal 11. Landscaping incorporates different aspects, enabling an integral design of urban ecosystems. This chapter defines the basic concepts of landscape, matrix, spots and ecological corridors which are essential for urban ecosystems to function. In this sense, it studies how fragmentation of green areas has repercussions on the isolation and function of city ecosystems. In order to do this, the study includes concepts of ecology that are applied to the study of ecosystems and constitute their physical and biological structure. This chapter also reviews the main methodologies used to study urban landscape, as well as the classification and identification of green areas, transects and multitemporal analysis without forgetting aspects related to the population's perception and their relationship with socio-cultural aspects. Finally, considering the peculiar characteristic of Lima city located in the desert and its cultural heritage, it highlights the importance of water in the design of the landscape.

Keywords

Landscaping · Landscape ecology · Ecological corridor · Fragmentation

It is necessary to know the principles of landscaping in order to design green areas in the city and make Lima a more inclusive and resilient city according to the Sustainable Development Goal 11. Landscaping incorporates different aspects, enabling an integral design of urban ecosystems. This chapter defines the basic concepts of landscape, matrix, spots and ecological corridors which are essential for urban ecosystems to function. In this sense, it studies how fragmentation of green areas has repercussions on the isolation and function of city ecosystems. In order to do this, the study includes concepts of ecology that are applied to the study of ecosystems and constitute their physical and biological structure. This chapter also reviews the main methodologies used to study urban landscape, as well as the classification and identification of green areas, transects and multitemporal analysis without forgetting aspects related to the population's perception and their relationship with socio-cultural aspects. Finally, considering the peculiar characteristic of Lima city located in the desert and its cultural heritage, it highlights the importance of water in the design of the landscape.

A. Sabogal, *Urban Ecology*, Sustainable Development Goals Series,
https://doi.org/10.1007/978-3-030-69905-5_2

Peru's earliest examples of landscaping include the *chulpas*, or pre-Incan tombs; many of these are located on the summits of hills, regarded by the ancients as divinities who took the dead. On these cold, windy and desolate places, small stone *chulpas* were built to house the deceased. Many of them are still intact today. Views of the landscape were much used in Inca architecture for territorial defense, and for contemplating the hill-gods that governed the world and human destiny. For this reason, the windows in Inca constructions always face towards the cliff-edge.

2.1 Landscapes Studies

Garden design was born as an artistic, almost religious, expression which begins with the creation of mausoleums as a representation of man's passage to the other world. Furthermore, it is a protected space, where we can be in contact with nature and creation. Garden design does not disregard humankind's domination over nature; gardens are created, first controlled and dominated by design. However, the need for parks and gardens arises only when cities expand, and people distance themselves from nature. Thus, landscaping is a human expression; a look towards nature and an intent to reproduce it while designing green spaces.

Landscape can be developed from different perspectives. On one hand, a natural perspective focused on the biological aspects of plant; on the other hand, a design based on an architectural perspective which does not consider plant growth or development. In this sense, city landscape design includes plant use. Patrick Geddes (1854–1932) tries to perceive and rediscover the geological aspects of landscape and combines design with social aspects; he applies methods of geography, biology, agronomy, geology, history and culture of space, developing an interdisciplinary methodology (Vercelloni and Vercelloni 2010). The study and design of the landscape is and ought to be an interdisciplinary science to achieve its objectives.

2.2 Urban Ecology Research Methodology

Urban ecology includes the use of plants; however, it has developed and led to the study not only of plants but also of all other species and their organization, shaping the city's living systems and ecosystems. In Lima, where water is scarce, there are ecosystems in which abiotic components reign and in which plants and animals cannot prosper. Thus, it is interesting to consider and appreciate the geomorphological aspects of the landscape.

The method of transects is often used to study urban ecology in the cities. Transects are linear spaces that enable the study of interconnections between ecosystems. This method is especially suitable in concentric cities (Haase and Nuissl 2010 cit. Endlicher 2012: 40). Transects make it possible to determine construction density, distance between green areas and their interconnections. Sukopp proposes the study of gradient analysis which defines three different spaces in the city: the city centre, the periphery and adjacent fields (Endlicher 2012: 40). In order to study and plan green spaces, initially, the type of space should be determined based on the use it is given and its ecological characteristics (Breuste 1996 cit. Endlicher 2012: 41). This permits to characterize and categorize spaces based on their components. There is a biodiversity gradient in the city that correlates with the distribution of species. In this regard, Sikorska and Sikorski (2017) try to determine an urban–rural gradient to measure the degree of disturbance, relating it to the distance from the city to the centre with respect to the periphery.

The urban–rural gradient can still be discerned on the city's southern and northern peripheries. Lima s situated in the Rímac Valley and has gradually expanded to the Chillón River, to the north of the city; and to the Lurín, in the south. Today, the Rimac and Chillón rivers are part of the city. Until the late twentieth century the Bocanegra and Oquendo estates could be found in the valley, north of Lima. Today, they also fall within the city. However, because of its proximity to the airport, part of the cropland of the old Bocanegra Estate, as well as its estate house, has been preserved. The Lurín River still lies beyond city boundaries and continues to play an important role in the production of crops and ornamental flora.

Kowarik (1992) defines four types of vegetation in the city. The first is composed by remains of natural ecosystems. The second, called cultural vegetation, collects anthropic elements and forms cultivated spaces within the city. It includes grasslands, meadows, agricultural fields or forest spaces; however, these spaces endure strong urban pressure. The third type of vegetation is the one found in gardens with planted and cultivated green areas. Finally, the fourth type of vegetation is formed by industrial or urban spaces where nature is reborn without the help of human beings (Kowarik 1992).

This classification is particularly interesting in Lima city where in spite of a very different reality with natural ecosystems formed by lower river basins and desertic ecosystems, we can still find urban-industrial areas, peripheral areas and densely populated areas in Lima centre, all intertwined with parks, squares and urban spaces, which make up a complex pattern. In all these spaces, new ecosystems are formed which use the water in the city and find protection from climate and high temperatures, forming islands of vegetation in a naturally desertic space with mostly introduced species.

To use the typology proposed by Kowarik (1992), Lima still has natural vegetation on its southern periphery, such as in the Pantanos de Villa Wildlife Park; and on its eastern periphery, along the sea cliffs. On its northern fringes, the city has squatter settlements that remain precarious, with incipient public services and high population density interspersed with some surviving cropland. A second type of vegetation, cultural vegetation, can be found in densely populated parts of the centre of Lima, with its colonial architecture, convents, plazas, crowded public areas, and narrow streets. Cultural vegetation has some natural elements but is subject to strong anthropogenic and urban pressure, such as that located along the banks of the Rímac and Chillón rivers. Species include wild-growing introduced plants such as ricinus (*Ricinus comunis*); cultivated fruit trees such as adocado (*Persea americana*) and banana (*Mussa x paradisiaca*); as well as native plants such as Humbolt's willow (*Salix humboldtiana*); sauco (*Sambucus peruviana*), and Peruvian pepper (*Schinus molle*). Also included in this category are the port of Callao and certain beaches such as Los Pescadores: parts of the city to where fish are landed for local consumption. These spaces host seagulls and pelicans in large numbers, as well as non-native fauna such as rats, cockroaches and many others. The only plants are a few trees planted by humans and other introduced species. The third type of vegetation is made up of gardens and green spaces, such as the large private yards and public parks in affluent parts of the city. Fourth and finally, there are revegetated urban-industrial spaces where nature regenerates without human assistance; examples include river mouths and beaches close to the city, where there are large numbers of seagulls. Many of these urban-industrial spaces are also located close to Callao. This category is typified by large avenues with very few trees, large working-class neighbourhoods containing residential complexes, and very few green spaces.

In order to study a city, urban spaces can be classified based on the use of space and the density of infrastructure determined by how compact construction structures are (Endlicher 2012: 43). This defines the distinctive patches that form the urban pattern. Urban density and uses as well as types of urban vegetation can be measured and defined by means of satellite photos (Lakes et al. 2011 cit. Endlicher 2012: 48). Therefore, we use the same methods as for landscape study in order to characterize urban space. These are the matrix, patches, mosaics, corridors and networks. The matrix is the dominant element or background of the landscape, whereas patches are disruptions which make up the mosaics and corridors are linear elements that connect the patches; finally, networks form the set of corridors (see Fig. 2.1).

Methods that use satellite photos can be of two types: the passive type which measures the

Fig. 2.1 Ecological corridor Costa Verde, Lima. *Author* Ana Sabogal

reflectance of the earth, and the active type which uses laser technology and includes estimation of time between the reflection and its detection (Wohlgemuth et al. 2019: 34). The passive optical sensors are optimal for measuring changes in aquatic ecosystems, they can detect its different colours and approximate vegetation heights; in both cases the Light Reach Detection program (or LIDAR) is used. These can be also used to calculate air pollution levels in the city or ultraviolet rays, while vegetation changes can be measured using the Normalized Vegetation Index (known as NDVI) (Wohlgemuth et al. 2019: 34). This method is also used for the study of clouds, which prevents visualizing the space clearly. Satellite photos of Lima are not always clear due to high cloud coverage and humidity levels.

Urbanization will influence and modify natural ecosystems by creating new ones, also introducing greater variety than in the natural ecosystems. To quantify the impacts of urbanization and characterize them, the most urbanized spaces can be compared with those less urbanized. In this way, we can define three scales, urban, exurban and non-urban and define with them the urbanization gradients (Ramalho and Hobbs 2012 cit. Francis and Chadwick 2013: 179). These will define the patches that may constitute mosaics formed by urban ecosystems. For the study of urban ecosystems, the impacts on biodiversity associated with biodiversity gradients will be quantified according to the type of native and non-native species, as well as the key species of each ecosystem. The classic parameters used for the study of biological communities such as physical and biological structure, dominance, relative abundance and biodiversity, will be essential to characterize ecosystems and define their proper management. To define the community, the Sorencen Similarity Index and the Shannon and Wiener Species Diversity Index may be used (Sabogal 2014). Moreover, studies of Lima that apply ecological methods are practically non-existent; most parameters are architectural and focus on present quantities of vegetation in parks and their surroundings.

For landscape assessment, the study of the landscape is practiced considering not only ecological aspects but also political and socio-economic ones (Zasada et al. 2017). They should be considered in the analysis since these are all interconnected and have a cascading effect (Zasada et al. 2017). This framework comprises a study based on territory without setting aside the social actors and therefore avoiding future conflicts; this way the process of territorial planning must be based on spatial dynamics and requires the inclusion of landscape assessment by the local population (Zasada et al. 2017). In this way, space planning will be more realistic and adequate to social needs. Citizen planning and social population studies provide for more realistic planning of space suited to social needs. Studies focusing on aspects of usage include Sabogal and Martinez (2015), which defines the typologies of Lima's parks; and Sabogal et al. (2019), which characterizes the use of parks in the district of Santiago de Surco in Lima. Also of interest, but taking a different approach, is Ludeña (2013), which considers berms, bridges and stairways to be public spaces given their high levels of usage.

Whereas landscape assessment is broad and is based on ecological aspects, the use of landscape introduces the value conferred by the population to this space based on the population's perception. Biecke-Matejak (2017) in his analysis of the Saussete forest park (Paris), sustains that Michel Corajoud distinguishes between system analysis and object analysis. System analysis involves spatial distribution whereas object analysis considers biodiversity. This means that when studying ecological corridors, on one hand we analyse the system of ecological corridor, and on the other hand, the objects that compose the corridor such as trees. When analysing the system, we should include the interrelations between its components. These are intertwined since habitats differ depending on the variety of species that settle in them. For an integral analysis, it is necessary to include the cultural value of the space as well as the value given by the population to each object (Biecke-Matejak 2017).

With regard to the typology of space, it is worth returning to Lima's Chinatown. Located in the heart of the capital, Chinatown makes intensive use of space for commerce, leisure and socialization; *Limeños* visit the area to buy, eat, stroll, and get their fortunes told. This is a high-transit intercultural space with multiple characteristics. Given the lack of plants, Chinatown is best regarded as a public space rather than a green area; and this being so, it exemplifies a key difference between the two categories. This landscape is entirely anthropogenic, lacking in vegetation but rich in culture and diversity.

The theoretical review about landscape's subject and object leads to the discussion regarding the quality of green areas, which different authors define and measure applying a diversity of parameters. To define the quality of parks and green spaces, on one hand, it is necessary to consider space size, ecological aspects and interconnections and, on the other hand, the use it is given. This subject can be undertaken, as several authors suggest, by decomposing green areas into its components in order to define quality from this point on and the use determines its functionality and consequently its quality (Sabogal et al. 2019). Photographs that show physical structure can help considerably to identify key species and indicator species. To establish a quality index, this should include aspects of species distribution such as the presence and frequency of birds, which correlates directly with the rooting process of a habitat in a city's ecosystem (Sikorska and Sikorski 2017). This emphasizes the continuity of the landscape and highlights the value of data compilation for a semiotic map (Gawyszewska 2017) which, together with an inventory of plants and sociological information, will allow the cultural aspects of the landscape to be revalued.

Conversely, the Pantanos de Villa Wildlife Park on the outskirts of the city is certainly a green area, but, given the lack of social interaction there, not a public space. Characterized by its large numbers of birds and other wildlife, this natural ecosystem hosts a large number of birds and other wildlife but very little sociocultural life or engagement.

The quality of green areas depends on various factors. If we focus on the technical parameters of greenhouse gas absorption, trees in the city are essential. To measure its effect, the *Urban tree database* (a global trees species database) considers the following parameters in the equation: leaf biomass, wood biomass, climate, species, growth data (McPherson et al. 2016). For this database, remote sensors are used to calculate by means of an allometric equation the foliar

Table 2.1 Main landscape function, own authorship

Main landscape functions	
Function	Parameter that define it
Social	Population use
Cultural	Historical value of the space
Ecological	Species and ecological functions
Aesthetic	Populations perception

biomass and the absorption of carbon, to later relate it to the socio-economic benefits derived (McPherson et al. 2016: 1). All this enables the calculation of the space required by each species to prevent conflict with infrastructure (McPherson et al. 2016: 1). However, changing management conditions in the city make it difficult to get accurate measurements (McPherson et al. 2016: 10). To minimize the margin of error, the model makes calculations under several types of climate, considering days of frost (McPherson et al. 2016: 6). This results in a generic database where the obtained data may vary when correlated with management species' parameters such as water, required space, fertilization, pruning, sanitary practices, among others. For this reason and due to the change in environmental conditions, it is more convenient to have an accurate assessment of trees for a period of time. Especially, if the database is used for a city where parameters are variable in each space studied. That is why, to be accurate it would be necessary to have a specific database applied to the equation for each case (McPherson et al. 2016: 4–5). This analysis is, however, very interesting and even though it should not be followed strictly, it consistently helps define management parameters and gives us light on how to apply them to reality.

In the study performed to the parks of Lima, criteria considered to measure the quality of the parks defined two types of parameters. First, parameters of components: infrastructure, vegetation, equipment and services. Second, parameters of system functionality: surface, frequency of visits by inhabitants, location with respect to the study area. Likewise, inhabitants' perception was also considered (Sabogal et al. 2019). Another study, also seeking to determine the quality of the parks in Lima, concluded that parks in mid- socioeconomic level districts are visited frequently for their high quality by inhabitants of low-income adjacent districts which only have 4.82 m^2/inhabitant (Tavera et al. 2018). These high-quality parks are better than those in higher income districts since middle sectors do not have private areas so more green areas are required (Tavera et al. 2018). This study was performed applying a mixed methodology in which parks were first selected based on the socioeconomic level of the district, to then perform an evaluation to determine the quality of the public space, considering the park's equipment. Finally, a survey was conducted among the visitors to include the population's perception of the park studied (Tavera et al. 2018).

We acknowledge 4 main landscape functions: social, based on the use that the population gives to the landscape, cultural, based on the value of the species or historical value of the space, natural or ecological that will depend on factors such as ecological corridors and the absorption of carbon dioxide or habitat diversity and aesthetics that depends on the population's perception (see Table 2.1).

In order to achieve a proper design and use of green spaces, social and cultural aspects must necessarily be integrated. The conception of green spaces is correlated with the perception societies have of these. In this way, by means of photography, Kowarik (2013) while studying the perception of nature, defines 4 models of perception for the city of Berlin correlated with the idea of free nature with no intervention: biodiversity patches, landscape, recreation and culturally representative parks. As a result, he determined that the park least valued by the population is the biodiversity patch park, while the most valued is the recreational park.

The landscape of the city can be classified by determining units of landscape, based on their characteristics and translating it onto a classification map. This landscape information can be organised according to use of space, ecosystems, pollution, carbon dioxide absorption and temperature. In this way, we can establish ranges based on landscape sensitivity, adaptive capacity and define its units with specific characteristics. To map the landscape, green paths and ecological corridors should be distinguished. Whereas, green paths have a social function, ecological corridors have ecological functions. Both functions can be represented on the map.

To define the quality of parks and green spaces we must consider a multiplicity of aspects, such as social, cultural, biological and aesthetics. The key for parks to have the acceptance of the population and, at the same time, allow the development of biodiversity is the effective combination of all variables. However, this is not always easy.

Applying theoretical concepts to Lima takes us to the Costa Verde (literally, “green coast”) ecological corridor (Fig. 2.1). This coastal strip possesses the four functions mentioned in Table 2.1. This is a much-visited space that hosts a range of social and cultural activities including sports, dance, and birthday events. With splendid views of the Pacific Ocean and many plants embellishing the space, the Costa Verde is a site of great aesthetic beauty, stretching several kilometres, as well as an important ecological corridor. To maximize this role, the space could be connected to a variety of Lima’s avenues, creating corridors running the breadth as well as the length of the city. It is vital that the Costa Verde be preserved as a public space for all *Limeños*, and not turned into a tourist site isolated from the city’s social and cultural reality. This corridor is described in greater detail in Chap. 4.

2.3 Multitemporal Analysis

Multi-temporal analysis is an important tool used for landscape study both for space monitoring and to define progress in terms of expansion and green area management. Furthermore, it is very useful to study changes of coverage and space. Since it enables a comparison of maps at different times. For this, satellite images are used such as Planet Scope (3 m resolution), Landsat 5 and 7 ETM and 8 OLI (30 m resolution). A subsequent supervised classification can be added using Geographic Information Systems (GIS), in order to recognize the spectral signatures of different classes, such as vegetation, bare soil, urban areas, among others. We can thus define the changes in city vegetation coverage in a period of time. Vegetation changes can be measured according to their size using the Standard Vegetation Index (NDVI), as mentioned in Sect. 2.2. This method is also used for the study of landscape elements.

2.4 Water as an Element of Landscape Design

Park and garden design involve the creation and management of ecosystems. To ensure that the ecosystems are in equilibrium, the elements that compose it are required to integrate with each other. In this process, each of its elements must be considered: air, water and soil. Due to the Spanish heritage and its Arab influence, water is an element that is highly present in Lima’s urban landscape, as well as in the beauty representations of the city’s imaginary. However, there are spaces in Lima that were not designed considering the movement of water, giving conditions for the presence of mosquitoes and generating health problems for inhabitants.

Water is an important focal point in landscape design, around which the design revolves. As Bachelard (1978: 54) mentions: “*The true eye of the earth is water.*” In this sense, water can lead to various interpretations of space. Whereas water in streams flows making us feel cheerful, calm and clean, water from fountains has the force to go against gravity, as opposed to nature. The usual sound of these joyful waters transmits the force of happiness and youth. Water is accompanied by the music produced by movement. Even before seeing water, we hear its music as a burlesque sound (Bachelard 1978: 282).

The world of water expresses the mystery of the unknown, especially if deep waters reflect the sky, as if surrounded by fog, characteristic of Lima at dawn. We also have this feeling when we observe the moon's reflection over the jungle rivers. This makes us think of something mysterious, but not of grief, because we know that they will help us navigate to reach our destination. On moonless nights it is harder to cross the river. The still and stagnant waters of a lake are silent waters. They belong to dark and mysterious nights without moon nor voices (Bachelard 1978: 287). Just as Narcissus looks at the reflection of the waters and looks beyond his reflection, Snow White also sees her future in water the water of the well.

The admiration we feel for nature can be expressed our design of the parks. For those of us who live in Lima or along the long Peruvian coast, the sea is a very important symbol that appears in dreams and inspires us. This is a space that should be embodied in the design of parks and gardens. The violent waters and the sea waves are part of the search of nature to participate in the natural world, full of challenges other than the human world, full of falsehood. This ambivalence can be used in the design, by incorporating the sea, its symbols and contradictions as part of it. Water with violent movement and waves gives us energy, conveys youth (Bachelard 1978); but it also transmits fear and respect, leading us to admire nature. Using it as part of a garden design can be interesting, at the same time charming and cheerful. Just like water, trees are also mysterious and can be a focal point for design. According to Jung, a tree is a maternal symbol just like water.

Old Lima was situated in the basin of the Rímac River, but the city now extents to the Chillón basin and the coastal desert. Until the later nineteenth century, water within Lima's city walls was drawn from the Cacaguasi springs in the hills that surround the city, and distributed to pools by way of enclosed guttering. Thus, vegetable gardens could exist within Lima, irrigated by canals such as the Huatica (Pacheco 2018: 11). Water from the Rímac was, and still is, also used for this purpose.

Water as an element of the public park is central to the Spanish cultural inheritance in Peru. The Parque de la Exposición, built in what was then the outskirts of Lima, utilized water from the Cacaguasi springs, allowing the production of crops. The Matamandinga Estate, which previously occupied that site, was supplied in the same way (Pacheco 2018: 13). Water today remains central to the city's imaginary. The Parque de la Reserva, known today as the Parque del Agua, was inaugurated in 1929 (Gastelumendi 1997: 84). The park was built during the Leguía presidency—characterized by the development of numerous public works, including the Parque de la Exposición—on a sixteen-hectare site belonging to the National Agricultural School and which was once the Santa Beatriz Estate (Orrego 2019). The renowned French architect Claudio Sahut was in charge of this highly ambitious and novel project, while José Sabogal, an exponent of the Peruvian indigenist trend, oversaw artistic creation and works execution; the park's ornamental *huaca* was designed by the latter. Today, the park's eight hectares contains thirteen water fountains, the tallest of which soars to eighty meters, illuminated by coloured lights (Orrego 2019). This park, whose design revolves around water, continues to enjoy enormous popularity among *Limeños*.

References

Bachelard G (1978) El agua y los sueños. FCE, México, p 291

Biecke-Matejak A (2017) Management and restitution of urban forest in landscape urban planning of urban and suburb areas. In: Congress presentation: problems of landscape protection and management in XXI century.

Organized by Warsaw University of Life Sciences, Polski klub ekologiszny

Endlicher W, Kraas F, Krajewski Ch (2012) Einführung in die Stadökologie. UTB, Stuttgart, p 272

Francis R, Chadwick M (2013) Urban ecosystems: understanding the Human Environment. Routledge, USA, p 220

Gastelumendi E (1997) Arquitectura paisajista. Ed. Vivero Italiano, p 160

Gawyszewska B (2017) Disappearing urban landscapes. Urban wastelands protection and management in contemporary cityscape of block-of-flats settlements in Warsaw. In: Congres presentation: problems of landscape protection and management in XXI century. Organized by Warsaw University of Life Sciences, Polski klub ekologiszny

Kowarik I (2013) Cities and wildness: a new perspective. Int J Wildness 19(3):32–36

Kowarik I (1992) Das Besondere der städtichen Flora und Vegetation. Deutscher Rat Für Landespfläge. Schriftenreihe Heft 61:33–47

Ludeña W (2013) Lima y Espacios Públicos, perfiles y estadística integrada 2010. PUCP, p 224

McPherson G, van Doorn N, Peper P (2016) Urban tree database and allometric equations. United States Department of Agriculture, p 86.www.fs.fed.us/psw/. Revised: 18/11/19

Orrego L (2019) Parque de la Reserva: 90 años. El Dominical de El Comercio. 17 de febrero del 2019

Pacheco JJ (2018) Parque de la Exposición: El Jardín de Lima. Municipalidad de Lima. Munilibro 5

Sabogal A (2014) Manual de ecología del Perú. Ed. Sociedad geográfica de Lima and Instituto de Ciencias de la Naturaleza, Territorio y Energías Renovables, Pontificia Universidad Católica del Perú de la PUCP, p 189

Sabogal A, Martinez M (2015) A study of ecological corridors in two quarters of Lima: Chorrillos and Villa El Salvador. In: Perspectives on global development and technology 14(2015):587–596

Sabogal A, Cuentas MA, Tavera T, Vargas F (2019) Espacios públicos: estudio del distrito de Santiago de Surco en Lima, Perú. In: Kawsaypacha N°3: 105–138

Sikorska D, Sikorski P (2017) How to measure the quality of green infrastructure for the city's effective spatial planning policy? In: Congres presentation: problems of landscape protection and management in XXI century. Organized by Warsaw University of Life Sciences, Polski klub ekologiszny

Tavera T, Sabogal A, Pastor P, Suarez O (2018) Importancia del estudio y análisis de la calidad cantidad y distribución espacial de los parques del distrito de Santiago de Surco en el contexto de Cambio Climático. Espacio Y Desarrollo 31 (2018):89–116

Vercelloni M, Vercelloni V (2010) Geschichte der Gartenkultur von der Antike bis heute. WBG, Darmstad, p 275

Wohlgemuth T, Jentsch A, Seidl R (eds) (2019) Störungsökologie. Utb Haupt Verlag, Gernany, p 396

Zasada I, Häfner K, Schaller L, van Zanten B, Lefebvre M, Malak-Rawlikowska A, Nikolv D, Rodriguez ER, Manrique R, Ungaro F, Zavalloni M, Delattre L, Piorr A, Kantelhardt J, Verburg P, Viaggi D (2017) A conceptual model to integrate the regional context in landscape policy, management and contribution to rural development: literature review and European case study evidence. Geoforum 82(2017):1–12

3 Botany for Landscapists

Abstract

This chapter analyses and describes the basic concepts for garden design focusing on conditions in Lima city. Taking the lead from the UN Sustainable Development Goal 11, there is a drive to promote the sustainable development of Lima. Attainment of this goal requires that parks be available to the entire population, based on principles of environmental justice in terms of access and distribution. If park design, species selection, and maintenance were handled appropriately, better parks would be available to all of the city's inhabitants with less efforts and fewer resources. This chapter explores the basic principles of design; describes the agronomic characteristics of plants, considering the functions of each part, the influence of environmental factors on growth, plant physiology in Lima's subtropical conditions, management of pests and diseases affecting ornamental species, and nutritional deficiencies; and concludes with the main rules for adequate maintenance of plants and parks of Lima city.

Keywords

Landscape design · Botany · Plant physiology · Plant · Pests

This chapter analyses and describes the basic concepts for garden design focusing on conditions in Lima city. Taking the lead from the UN Sustainable Development Goal 11, there is a drive to promote the sustainable development of Lima. Attainment of this goal requires that parks be available to the entire population, based on principles of environmental justice in terms of access and distribution. If park design, species selection, and maintenance were handled appropriately, better parks would be available to all of the city's inhabitants with less efforts and fewer resources.

This chapter explores the basic principles of design; describes the agronomic characteristics of plants, considering the functions of each part, the influence of environmental factors on growth, plant physiology in Lima's subtropical conditions, management of pests and diseases affecting ornamental species, and nutritional deficiencies; and concludes with the main rules for adequate maintenance of plants and parks of Lima city.

Lima, where a third of the population of Peru is concentrated, is today marked by vast social differences and environmental concerns in the more impoverished municipalities, necessitating a rethink of the city. The lack of public spaces and colossal environmental pollution in cities are issues that must be tackled if the quality of life of inhabitants is to be improved. Only in this way can cities and human settlements become resilient and sustainable. Parks must play an environmental and social role that compliments city development. This chapter seeks to address the question of achieving sustainable green areas that

A. Sabogal, *Urban Ecology*, Sustainable Development Goals Series,
https://doi.org/10.1007/978-3-030-69905-5_3

help to improve the quality of life of populations. In the context of Lima, where water scarcity and climate change are forcing us to consider the resilience of the city, this line of enquiry is a necessity. Thus, the chapter reviews and proposes agronomic solutions. The botanical species outlined in this chapter are ones whose growth is possible in Lima.

3.1 Basic Principles of Garden Design

To design a garden, not only do we need to follow certain basic principles of design, applicable to any design but we also should think about the fourth dimension. This means we need to consider that a garden will change with time and it should stay in perfect balance while plants are growing. This is possible with the right nutrients and care so that plants grow and develop properly. A garden won't be beautiful if the plants are not properly nourished or if they are mistreated.

Although Lima was founded in the river Rimac basin with a riverside mountain ecosystem, nowadays it spreads across three basins: Rimac river, Chillon river and Lurin river. However, the coastal Peruvian desertic ecosystem is in the midst. Thus, Lima is built on two ecosystems: desertic and riverside mountain. Without maintenance and watering, gardens in Lima would cease to exist. They are dependent on a watering system. This implies that practically every plant has been introduced, usually, from another continent, specially Europe after the Spanish conquest, sometimes from another part of the country.

William Gilpin (1724–1804) is one of the first authors to distinguish between parks and gardens. This author proposes the garden to be designed as a dynamic, three-dimensional picture leading to what we now call the fourth dimension (Kluckert 2000: 393). It is not only important to know how plants should be planted, but also how the plant will grow and continue beautiful with the passing of time. The discussion to define gardens and parks continues in the twenty-first century. Shirley Hibberd (1825–1890) starts off by saying that each person has an aesthetic vision of the world, so she suggests that each person can design their own garden based on their concept of art (Kluckert 2000: 397). This led to a new discussion about the gardens, now called biogardens. According to Hibbert, flowers are very important in a stressful industrialized city since they compensate for the hectic life away from nature. Flowers give people another vision of space, especially, in the case of people with low incomes who do not have many options for entertainment (Kluckert 2000: 397). Christian Hirschfeld, landscapist who published the Theory of Garden Art at the end of the eighteenth century developed the idea that inhabitants of a city need gardens for physical and mental recreation (Kluckert 2000: 406).

We will see that especially in the case of the parks and gardens in Lima, we can apply the Spanish saying: "meterse en un jardín" ("*tresspassing into a garden*"), which means "*to ramble in a speech or in a theatrical discourse or to get involved in a difficult situation*", (translated from the Diccionario de la Real Academia Española 2019), since the charm of gardens is that they are complicated. The lack of fertile soil, water scarcity, sociocultural differences, and a shortage of resources in municipalities make the planning and maintenance of green areas in Lima especially complicated.

Among the essential basic principles of design, we have form, structure, texture, colour, light and shade, focal point and the fourth dimension. As follows we will discuss each one of these principles relating them with utility and comfort, economy and beauty of the garden, considering that in garden design we have vegetation and architectonic elements which should be in harmony with each other.

Indeed, the lack of funds earmarked for plant maintenance is a problem in the city, and if the principles dealt with here are not taken into account and incorporated into design, green area maintenance will remain difficult. The abovementioned issues, added to the dearth of planning around green-space maintenance, have contributed to an imbalance in the species and other components that make up public spaces.

3.1.1 Form

Based on growth, plants have different forms. We can analyse plants individually or as a group of plants of a species used as a set in the design of a garden. A compact group of plants has a form just like individual plants do. Plant form defines structure based on the way the plants grow. They can do so as trees, shrubs, groundcover and vines. Vines can be shaped with pruning and management. Thus, they can be trailing or scandent.

The concept of form also includes architectonic elements, such as paths or buildings. The form of garden elements, both of vegetation or non-vegetation, can be symmetrical when geometrical shapes are used or asymmetrical (see Fig. 3.1).

Trees can be classified based on the way they grow in the following way:

- Triangular, like pines and cypresses. (*Cupresus sp.*)
- Spherical, like the ficus *Ficus bejamina* o *Ficus pandurata* (*Ficus pandurata*)
- Semicircular, like the Poinciana (*Delonix regia*), or molle serrano (*Schinus molle*).
- Cubic, like the rubber tree (*Ficus elástica*) and the European oak (*Quercus robur*)
- Columnar, like eucalyptus (*Eucalyptus camaldulencis* o *Eucalyptus globulus*).
- Abstract, like palms, bamboos or sago palms (*Cycas revoluta*).

Flowers can have the following shapes (Muñoz 1979):

- Spike, like sage (*Salvia splendens*) or the snapdragons (*Antirrhinum majus*).
- Round, like daisies (*Leucanthemum vulgare*) or chrysanthemums (*Chrysanthemum morifolium*).
- Intermediate, like irises (*Iris germanica*), cyclamens (*Cyclamen persicum*) or orchids (*Cattleya sp. o Phlalaenopsis sp.*).

Even though most plants are not geometrical, they can become so by pruning and handling. The form of design is specially determined by the paths which can be geometrical, symmetrical or sinuous and asymmetrical. This is essential to define the type of design. If a classical design is desired, geometrical lines are suitable. On the other hand, if a naturist design is preferred, priority should be given to asymmetrical shapes for paths and the other elements of the design including plants.

The use of geometric or abstract forms depends on fashion and the way of thinking of an age.

Geometric designs are common among native and aboriginal cultures like the beautiful designs of the shipibos or those of the aboriginal Australians. Curiously in the twentieth century with constructivism and futurism there is a return to geometrical designs. That is why the twentieth century English landscapist Lawrence Johnson, among others, proposes to use geometry again in garden design by pruning trees to maintain geometry. He believes gardens should have thematic subdivisions and continue to be marked by visual axles to organize the park's design. The design should be integrated with the

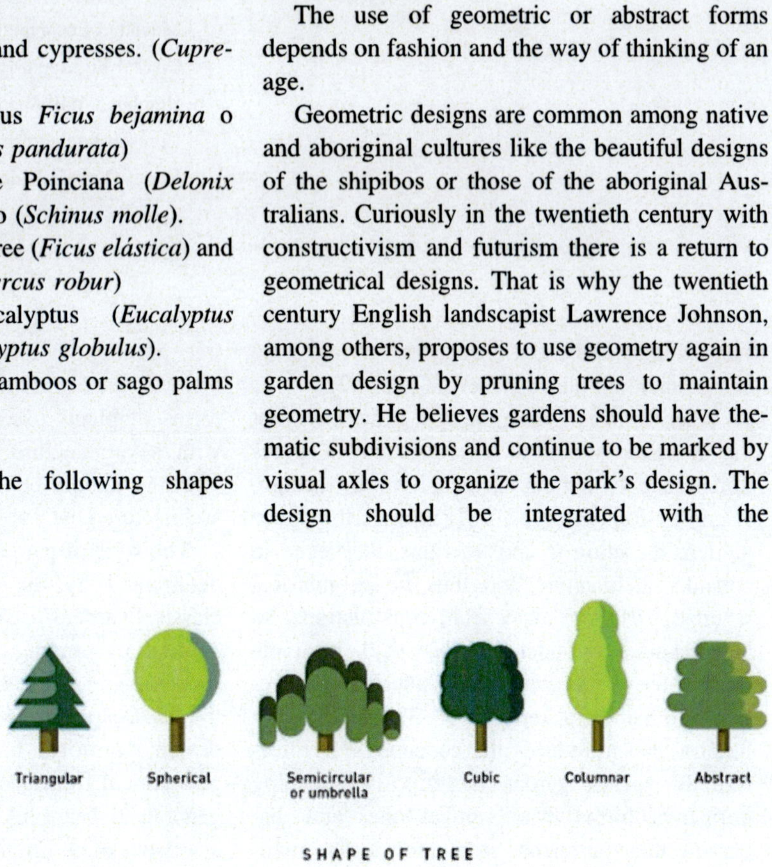

Fig. 3.1 Shape of trees, own authorship (*Illustration* Juan Pablo Bruno. *Source* Muñoz 1979)

Table 3.1 Types of trees based on tops, own authorship, illustration Juan Pablo Bruno

Types of trees based on their tops	
Triangular top	All type of conifers like:
	Radiata pine (*Pinus radiata*)
	Chilean Pine (*Araucaria araucana*)
	Thuja (*Thuja occidentalis*)
Spherical top	Apple *(Malus domestica)*
	Pear (*Pyrus communis*)
	Weeping Cypress (*Cupressus funebris*)
	Avocado (*Persea americana*)
	Mango (*Mangifera indica*)
Semi-circular or umbrella top	Weeping Willow (*Salix babylonica*)
	Poinciana (*Delonix regia*)
	Mountain Molle (Schinus molle)
Cubic	Cedar (*Cedrela odorata*)
	European Oak (*Quercus robur*)
	Rubber (*Ficus elastica*)
	Ficus pandurata (*Ficus pandurata*)
Columnar	Black poplar (*Populus nigra*)
	Mountain Eucalyptus (*Eucalyptus globulus*)
	Coastal Eucalyptus (*Eucalyptus camaldulensis*)
	Humboldt Willow (*Salix humboldtiana*)
	Crimson bottlebrush (*Callistemon citrinus*)
	Italian Cypress (*Cupressus sempervirens*)
Abstract	Sago Palms (*Cycas revoluta and Cycas circinalis*)
	Palms of all species

surroundings and use a lot of grass to make the design more ample (Kluckert 2000: 479).

When form is used in design, it defines what the designer wants to express, and it gives importance to certain elements of the design. Humphry Repton (1752–1818) thinks that since gardens are close to constructions, they need to consider architecture and thus be geometrical whereas parks are away from constructions, so they should consider the artistical principle marked by nature, without considering architectonic forms (Kluckert 2000: 394). Humphry Repton also introduces the concept of architecture as part of garden design, distinguishing horizontal forms from vertical ones and discussing their pertinence in relation to the architectonic constructions and the combination of tree pruning with architecture (Kluckert 2000: 394). To stress architecture and vegetation, the forms of plants used in design should contrast with the architecture. To camouflage it, the forms of plants should be similar to those used in the architecture (see Table 3.1).

The style of garden, to a great extent, will be determined by the choice of form. Whereas classical gardens use geometrical forms and great amount of pruning to keep the plant's form, modern gardens use natural abstract forms. Since the landscapists introduced the English garden design, priority has been given to non-geometrical forms both for plants and design in general. A beautiful design based on the characteristics of an English garden is the Babelsberg castle garden outside Berlin. Prince Herman von

Pflücker-Muscau applies this style and determines that the space should represent nature in its natural state (Endlicher 2012: 194–196).

The Babelsberg castle garden is the result of this naturalistic vision of garden design. William Kent (1664–1700) was the designer and executed the works together with the prince. He looks to break with the dependency on garden geometry, creating spaces that imitate nature with spaces for entertainment and picnic that form aesthetic landscapes. With the intervention of prince von Pflücker-Muscau, a rose garden, a water music garden, among other beautiful designs are integrated into the garden. The landscapist Peter Joseph Lenné also participates in the design by outlining paths and focusing on vistas from the castle to Potsdam (Sademann and Kilimann 2017). In this sense, he is clearly a landscapist and takes us back to the Venetian vistas so distinctive of paintings. We also notice a new element that involves both the space designed and the surroundings of the design (see Fig. 3.2). The design includes Greek temples, artificial ruins, grouts, hermitages, Chinese pagodas, water mirrors, still-water lagoons and streams which help outline a natural space (Endlicher 2012: 194).

The interpretation of form depends on culture and on the interpretation of space. Hischfeld in "*Theory of Garden Art*" (1778) (Kluckert 2000: 406) proposes to classify gardens according to the emotion they give. Ruins play an important role in this. Gardens like Glinicke have an important collection of ruins brought from different parts of the world. Amidst the distant views stand the Klein Glienicke castles (see Fig. 3.3), designed in the 1816, from where we can see the Hunting Pavilion Façade designed by architect Shinkel in 1824 (Kluckert 2000: 429). Another design by Leneé is the biggest park in Berlin called Tiergarten and Friedrichsfelde castle today called Berlin Tierpark (Endlicher 2012: 196). All these designs are filled with nature and have a lot of landscape views. The design of greenhouses, fashionable in the seventeenth and eighteenth century, is also important. These were called "*orangerie*" because citrus trees from

Fig. 3.2 Babelsberg castle garden, Berlin. *Author* Ana Sabogal

Fig. 3.3 Glienicke castle. *Author* Ana Sabogal

warmer climates were planted there to protect them from winter. This tendency was stronger in the nineteenth century with the development of the iron industry used in winter gardens and of great importance in garden design (Kluckert 2000: 457). The most famous of these is the Kew Garden designed by architects Decimus Buton and Richard Turner between 1845 and 1847 (Kluckert 2000: 457). Another important work is the *Orangerie* es Sanssousi in Podsdam, Germany where there are terraced steps to house plants that are moved to the *orangerie* in winter. In addition to the citrus plants la *orangerie* has exotic palms brought from different parts of the world. The development of the art of iron building enables the design of bigger spaces, such as the greenhouse in Champs-Elysees which holds meetings, balls and all type of social events.

The architecture of the gardens of Lima has long conformed to universal trends; thus, classical Lima used classical design forms, while the gardens of the twentieth century are more republican in style, utilizing large trees with extensive canopies.

3.1.2 Structure

Structure is determined by the dominant forms in a garden. We will apply the meaning given by the "Diccionario de la Real Academia Española" (Dictionary of the Royal Academy of Spanish) which defines structure as "*the distribution and order of the important parts of a building*". Humphry Repton (1752–1818) introduces this concept to design, distinguishing horizontal forms from vertical ones and discussing pertinence in relation to architectonic construction as well as the use of pruning to create and combine trees with architecture (Kluckert 2000: 394). This concept applies the term from an architectural point of view. Forms determine the height of the space designed, which is the vertical limit of our design. This term is also found in ecology to limit the space used by the species that are part of the ecosystem, which is called Physical Structure, different from Biological Structure which refers to plant composition.

Forms which determine structure, are trees, shrubs and groundcover. Vines can determine different structures depending on pruning and handling. These can be trailing or scandent so they will be part of the vertical or horizontal structure respectively. Structure guides design since it will define all the details. In a garden there must be a balance between the vertical forms defined by the trees and the horizontal forms defined by the plant beds. Structure can be tall when composed by trees, medium when formed by shrubs or low when formed by groundcover. Generically, we can speak of vertical or horizontal structures. When a garden has a vertical structure, there are more trees and shrubs. On the other hand, when its horizontal, there are more groundcovers and flowerbeds. For garden design, we should consider that in architecture, trees are equivalent to roofs, shrubs and vines to walls, plants used to form edges are like mouldings and groundcovers are like floors.

Le Crobusier (1887–1965) develops the idea that green is the background of architecture and he applies it in his design of the Contemporary Village, which is surrounded by a park crowned with a green roof and hanging gardens. In this sense, the roof is an architectonic frame that limits the design separating the sky from the design (Vercelloni and Vercelloni 2010: 242).

The design of parks in Lima commonly evince horizontal structures, sizable lawns dotted with some trees, and a lack of practical bush usage.

3.1.3 Texture

Garden texture is determined by leaf size, form and pubescence as well as trunk bark. Texture refers to the details of the elements, as defined by the Dictionary of the Royal Academy of Spanish, texture is "*the disposition and order of thread in fabric*". In garden design, texture is determined by the elements in the leaves and the trunk or both.

Types of texture for plants and garden elements are:

- fine
- medium
- thick.

A garden with a balance between different textures will be nice and pleasant. On the other hand, a disarray of textures will result in a disturbing and untidy garden.

Texture also refers to the way leaves are arranged on a plant and their density. Big leaves like the Ficus pandurata (*Ficus pandurate*) give intense shade, marking great difference between light and shade. Thin leaves like those of the Mexican creeper or coral vine (*Antigonon leptopus*) or the cananga tree (*Cananga odorata*) let the light go through forming a tulle. Dense and dark leaves like the conifers absorb light creating dark colours. Based on texture, plants can be dense or sparse, which will define the transparency of the plant and the quality of the light that will go through. Shadows will make the plants appear to be bigger. So, there are transparent plants and dense, dark plants that do not let the light go through.

The traditional use of overlays was central to the gardens of classical Lima, in which creepers featured largely, forming pergolas. Today, the parks of Lima have abandoned overlays, prioritizing instead thick textures and dense plants.

3.1.4 Colour

The basic principles of colour for a garden are the same as for painting. There are warm colours like red and yellow and their combinations as well as cold colours like blue and green and their different combinations.

Warm colours make spaces smaller and give the ambience warmth and happiness. Cold colours give depth to a space but also melancholy. These colours are often used in old classical gardens; they remind us of the passing of generations. Green, which is the main colour in a garden, is usually cold. However, depending on the colour it is mixed with, it could be warm. Whereas the dark green of dense trees in the amazon rainforest is cold since blue is prevalent, lemon green in spring leaves is a warm colour since the prevalent colour is yellow.

In design one can play with colours. We cannot forget that leaf colour changes with climate. In spring and at the beginning of summer, plants around the world are lighter, whereas at the end of summer and in winter leaves become darker. In a place like Lima, where seasons do not change a lot, we can still play with changing colours by choosing plants that flower in different seasons creating patches of colour that change the focal point throughout the year. For example, using the flamevine (*Pyrostegia ignea*), one of the few vines that flowers in winter or the Mexican creeper or coral vine (*Antigonon leptopus*) which flowers in spring–summer or the honeysuckle (*Lonicera periclymenum*), which flowers at the end of summer, the garden will be different in each season of the year.

Monet Garden in Giverny is one of the best depictions of colour in a garden and it is a transition to abstract art (Kluckert 2000: 465). Roberto Burle Marx (1909–1994) also proposes the use of plants as strokes of colour, going back to the idea of the Monet Garden in Giverny (1890).

While Lima during the viceroyalty was known for its use of pastel colours, republican Lima switched to cold, sharp colours, and green reigns in the parks of Lima today.

3.1.5 Symmetry

Symmetry is a highly discussed issue in landscape design; whereas classics prefer symmetry, modern gardens highlights the natural form of plants, letting them grow freely. In his book, *Rustic Symmetry*, Stephen Switzer (1682–1745) rejects symmetry so common in French gardens (Kluckert 2000; 392). This style causes art gardens to distance from gardening. Thomas Whatley in his book *Observations of modern gardening illustrated by descriptions*, published in 1770, considers that garden art should centre on an artistic vision rather than on gardening (Kluckert 2000: 392).

It is important to point out that initially gardens were developed by forest engineers, following the landscape school, whereas currently most schools are developed in line with the

Fig. 3.4 View from Babelsberg castle. *Author* Ana Sabogal

architectonic school. The agronomical vision of a garden so present in the middle ages with the herbalist garden has been disregarded. In the twentieth century, the vision of a garden is from the architectonic point of view. The utilitarian function of the plant is overlooked, and it is used only for design. Sight of the fourth dimension is lost. Therefore, the fact that a garden changes with time and the passing of the seasons is not considered.

Babelsberg park, outside Berlin, still integrates both disciplines, developing a garden with an interdisciplinary vision. We see how in the twentieth, the interdisciplinary vision is broken resulting in the division of disciplines typical of the twentieth century. It is at this point that the vision breaks drastically.

When we analyse Babelsberg park mentioned above, we can see it follows the vision of forestry through landscaping architecture, integrating the surroundings to the design. This park was designed by 3 intellectuals of the time. First, by William Kent (1664–1700) who designed and executed the work and later prince Herman von Pflücker-Muscau who followed the style of the English garden, but combined it with previous styles like the French by building pergolas, rose gardens and mazes integrating them to the English style which depicts nature in its natural form and finally Peter Joseph Lenné, who integrates the surroundings by designing views from the garden (see Fig. 3.4).

Symmetry promotes good arrangement of space and ease of management; parks of uniform symmetry and wide berms are easier to manage, facilitating the use of machinery for maintenance. It is also easier to plan which plants to use and control resources. Thus, most parks in Lima prioritize symmetry in their gardens.

3.1.6 Light and Shadow

Garden light and shadow are determined by form and size of leaves and plants as well as density of plants. Form combines with the definition given by light and shadow. Let's remember that the bigger the shadow, the more density and size the tree will appear to have. Light and shadow also define the size and form of the plant.

The use of different hues of the same colour with a variety of intensities, gives a similar result to that produced in black and white photography, enhancing plant form. Dense plants, corrugated bark or big leaves project a shadow and cause a contrast of light and shadows in a garden modifying the appearance of size and form. Thus, the landscape has more depth, melancholy and sadness. Whereas, if we want to highlight colour, we should combine colours emphasizing warm ones which will give light and joy to the design.

Playing with light and shadow enables us to change the dimension of space. When dark colours are placed in back and light colours in front, the space appears to be smaller. On the other hand, when light colours are placed in back, the space appears to be bigger. Shadows give depth and put emphasis on form; Conversely, light hides forms and puts emphasis on colour. In this way, playing with light and shadow to a great extent will define the garden's character.

As with other aspects of design, but with even greater emphasis, the use of shade has followed historical trends; the gardens during the period of Peruvian romanticism in the late nineteenth century made heavy use of it, while the gardens of the viceroyalty were more illuminated. But the concept is not currently employed in the design of parks in Lima. The selection of plants for parks is determined by ease of propagation, with priority given to species that can be propagated via cuttings.

3.1.7 Fourth Dimension

Landscape design is very close to painting, especially when it integrates the surroundings; watercolour techniques are used in the sense that distant space blurs at the sight so that the focus is on the details of what is near, yet, there is still an uncertainty about what is far and unknown. The fourth dimension also includes the vision of design size. William Gilpin (1724–1804) is one of the first authors that distinguishes gardens from parks. He proposes to design a garden as if it were a dynamic, four-dimensional painting, concept that has led to today's fourth dimension (Kluckert 2000: 393).

Since garden design uses plants, time should be integrated to the design. The challenge is to accomplish a nice garden from the beginning that will remain beautiful several decades. The beauty of a garden to a great extent is owed to the beauty of the plants during their whole Lifecycle, so that we can appreciate them in each stage and discover their beauty. When a tree is properly planted, it will probably outlive us, always in the same spot, it will be part of the history of the area and whenever we plant a new tree, we will think about future generations. With the proper nutrients, the tree will have a dignified life and remain standing in silence in contemplation of our past and future lives.

For this reason, when we design a garden, to ensure the growth of each plant and the integrity of the garden, we should consider the adult size of the trees and plants and place them in the blueprint. In this way, we will make the proper choice of trees and shrubs. While the trees are growing, we can include plants of temporary growth to fill up the empty spaces. The challenge is to make a series of designs throughout time.

The garden of the Liebermann Villa, the summer residence of painter Max Liebermann (1847–1935) located at the shore of lake Wannsee, once the outskirts of Berlin, follows this vision. It is certainly a romantic vision that helped the painter create a great number of beautiful paintings. The garden and bio-orchard are integrated to one vision with beautiful colourful flowers and views of the lake and groves. It was designed by Max Liebermann and Alfred Lichtwark, landscapist and director of the art salon Hamburg. The garden accentuates colours, plays with light and shadows and enhances the beauty of the space (see Fig. 3.5).

Trees are no longer favoured in the design of Lima's parks and streets, because they compete with electricity and telephone cables distributed throughout the city, and because their roots, especially those of tropical species, tend to disturb the sidewalks.

3.1.8 Focal Point

A focal point is the first place we look at when we contemplate a space, hence, it should be carefully chosen. Isolated plants could be focal points but sets of plants or groves could also be

Fig. 3.5 Lieberman Villa. *Author* Ana Sabogal

Fig. 3.6 Pole to ensure that a tree will be straight. *Author* Ana Sabogal

so. In the latter no plant would stand out individually. Too many focal points make for a messy and restless garden rather than a peaceful one at sight. Focal points should be determined based on the size and style of the garden and will be the axle of the design.

When we define a focal point, we should consider the architecture of the buildings and combine the design of the garden with it. A very big tree makes a house look smaller and vice versa. On the other hand, if the tree and building are the same size, the tree can cover the architecture of the building or compete with it, or it can enhance the building if it has different texture, size and form.

There are gardens with sub-divided spaces. In this case, several focal points are used, one in each sub-space. A focal point could be out of sight and appear as one approaches. This is the case of the Italian gardens with their winding paths that hide focal points. We can talk about the "surprise effect" which is caused by the impact of discovering the hidden focal point. In other cases, such as the French gardens, the focal points are very few, or even just one guiding the groves. In an English garden the focal point is usually a natural element, such as a tree. This garden design is ample, good for taking a stroll different from the French garden where the focal point is the priority of the landscape design, in these gardens the focal point can be a view outside the design of the garden.

It is also William Gilpin, who proposes to handle garden design as a painting in which visual creation is directed to focal points; He proposes the use of avenues with trees or what we currently call visual groves (Kluckert 2000: 393).

Most parks in Lima have a central plaza that acts as a clear focal point to which all the footpaths lead. This is not unlike the old plazas of central Lima with fountains at the heart, an example of the Spanish design legacy (see 4.1.1).

3.1.9 Garden Usefulness and Comfort

When designing a garden, it is important to combine art and comfort. A garden should be comfortable for those who are going to enjoy it, both in the case of public spaces and private gardens. If a public space is difficult to access, it will not be prioritized by citizens, which is also the case for insecure places or those that lack facilities. Another important factor in choosing a park is interconnection between parks in order to enable long walks and stroll.

Since the end of the nineteenth century when parks began to be associated with health, strolls and fresh air, parks have been seen like a health necessity for the citizens (see Chap. 1). The German doctor Leberecht Migge (1881–1935) introduces the concept of park usefulness affirming that they should provide citizens with entertainment and health and therefore should have space for amusement and sports (Kluckert 2000: 476). At the same time, Martin Wagner in his book "*Das sanitäre Grün der Städte*" (The Sanitary Green of Cities), published in 1915, supports the air purifying role of parks in the city (Kluckert 2000: 477).

If the garden is big enough, there could be several spaces for different purposes. For example, there could be an area to sit and enjoy the beauty of the garden or read under the shade of a tree and separate spaces for children to play hide and seek.

To satisfy the utilitarian role of a park the plants chosen should be suitable for the people that will use it. Furthermore, the role assigned to the park should also be considered. For example, if it is a school garden or a public park, plants cannot be poisonous like the lantana flower (*Lantana cámara*); If it is a garden for senior citizens or a public park, there cannot be trees or vines with slippery leaves or flowers like the Kapok (*Ceiba trichistandra*), the jacaranda (*Jacaranda mimosifolia*) or the flamevine (*Pyrostegia ignea*). Extremely fragrant flowers like angel's trumpets (*Brugmancia arborea*) or white jasmines (*Jasminum officinale*) could be a problem for people with allergies or overpowering when planted in a hospital where people have to stay for some days, but in a big park they could be very pleasant.

Although the use of gardens is important in the design, the other components should not be overlooked. Most municipal gardens are of a simple design, with walkways that form a cross, and a statue of sculpture placed in the center. This design is very practical from a spatial point of view, but if other elements are not taken into account it can prove inelegant.

3.1.10 Garden Economy

Plant selection is also important for garden sustainability. Plant should all have the same requirements of water and soil but different size crowns and roots so that vertical and horizontal spaces in the air and underground can be made the most of. Water and soil requirements can be easily determined if the plants are native from similar ecosystems.

It is important to consider that there is a direct relationship between the size of the crown and that of the roots which are one third the size of the crown, both in depth and total weight of the roots, unless their structure is modified through pruning or maintenance. In the city, trees are given a formation pruning so that they do not get too big and prevent the crown from affecting the cables or the roots from breaking pipes. This should be done before the tree is one year old by eliminating the terminal bud both from the root and the crown. Without this treatment a tree will grow taller, even if it is pruned subsequently (see point 3.2.1).

Another aspect to consider in design is space economy. Plants should be distributed based on the design, always considering the fourth dimension. If plants are placed too close to each other, when they reach their final size, they will take up the whole garden, break sidewalks or block pedestrians forcing pruning and causing structural problems. This affects the tree's health and stability and consequently increases its risk of falling.

The park economy should take into account costs and plan them accordingly. Most of Lima's parks fail to adequately consider this dimension, leading to the deterioration of parks after installation due to a lack of investment in maintenance. This only creates higher maintenance costs in the longer term. It is a problem that is most evident in low-income districts deprived of funds for ongoing maintenance.

3.1.11 Garden Beauty

When designing a garden, it is difficult to achieve the desired effect. If harmony is what the design is looking for, all the elements in the garden should combine. However, we cannot forget that a garden without some contrasting elements is monotonous. Therefore, some garden components should be opposed. This is achieved by contrasting plant textures or other elements in the garden such as plants and architecture. However, when contrasting elements, we should make sure not to include more focal points than originally designed which could cause discord. To achieve balance in a garden we should consider a balance in the number of each element. The biggest mistake is to place a few of each instead of a group of each element. In this way, the different elements of the design will be in balance as well as the design itself (see Table 3.2). For example, planting one sage (*Salagevia splendens*) in the middle of a field of grass is different from planting a bed of sage *(Salvia splendens)*.

We cannot forget that the final beauty of the garden will be determined by the group of landscaping elements in combination with the useful elements of the garden. A garden that is not useful and requires too much time and money to maintain is not sustainable, neither as a private garden nor as a local park. A garden that will be enjoyed mainly by children with no open space for running or playing hide and seek, will end up mistreated with children playing in the plants and flower beds.

Park beauty must be synchronized with the customs and culture of the population. In Lima, children need space to play soccer, and failure to factor this into park design can create conflict between aesthetics and leisure requirements. Indeed, parks full of "keep off the grass signs"—a common sight in Lima—are scarcely things of beauty.

3.2 Design with Plants

In this section, plants are described in terms of their forms of growth. The species and agronomic practices discussed are those most frequently employed for the parks of Lima.

3.2.1 Trees

The role of trees is very important: They give shade, decrease city noise, filter pollution, absorb carbon dioxide and give shelter to great number of birds, squirrels and insects. Trees have a variety of habitats for decades.

Trees are the vertical dimension of a garden. They are the roof of our design. They can be used as focal points to break a monotonous space or form groves leading to a piece of architecture or art. Whereas trees used as focal points should be

Table 3.2 Garden design contrasts. Own authorship

Garden design contrasts	
Wavy topography	Flat topography
Stones	Flowing water
Inanimate objects: stone, sand or floors of different colours	Animate objects: plants and animals
Sunny areas	Shady areas
Plants with flowers	Plants with decorating leaves
Sand	Red clay
Still water	Water fountain
Winding path	Focal point

eye-catching, either because of their size, form or colour, trees used for groves should be tall and straight to guide pedestrian. Furthermore, grove trees should not compete with one another, so they should be of the same species.

When a tree is between 1/3 and 1/2 of its final diameter it has reached its maximum height; from that point on it will only become wider (Smith and Smith 2007). Trunk width reveals the age of the tree. A standard method used to determine the diameter of a tree trunk is at the height of an adult's breast (DBH). This is required to discover the age of the tree as well to know the tree's condition in order to give it timely care and maintenance. However, height is also important to determine the condition of the tree. When the main trunk of a tree is damaged at an early age, it never reaches great height, since branches begin to grow sideways and stop vertical growth. This happens due to a hormonal change caused by the loss of hormones in the tip of the tree. Since trees in the city should not be so big, modifying their size benefits the city. That is why trees are submitted to a formative pruning that, as mentioned before, is done by cutting their terminal bud. Trees should be pruned before they are one year old.

In the city, trees are often planted inadequately and are submitted to too much stress. These circumstances are for example: too little space, bad soil, pollution and frequently, lack of water or over pruning. Another problem is that very few species of trees are used in cities (Pauleit et al. 2002 cit. Francis and Chadwick 2013: 104). All of this provokes pests and disease to spread so trees are pruned too often and become too sensitive.

For this reason, when choosing a tree, size and form of both the tree and its roots should be considered as well as the changes it will go through in form and structure in time. The bark texture and colour should combine with the colours in the garden. For a tree to be useful, its shadow should project onto the windows or garden, thus, this projection should be considered when selecting their distribution. This is also the case for the groundcover chosen to go under the tee. We can never forget that trees will modify as the crown keeps growing.

When selecting a tree, the colour of the foliage and flowers should be considered as well as how long the tree will have flowers. A tree could have beautiful flowers but only bloom for a couple weeks a year, or with inappropriate climate, never bloom. Such is the case of the Poinciana (*Delonix regia*), that will bloom only when it is warm enough and has not bloomed in Lima for some years whereas in Ica the tree is covered by lavish clusters of majestic flowers every year.

Depending on the tree's natural climate, it may shed leaves during the cold weather and go into what is called the dormant period. Dormancy is stimulated by the decrease in the hours of light and temperature with the arrival of autumn. Both environmental conditions regulate the activity of the plant hormone called abscisic acid (ABA). These two factors cause the decrease in the physiological activity of the plant that leads to hydric stress, both when it is cold and when there is a draught. At this point, the plant hormone ABA acts causing the stomatal opening to close and decreases plant's nutrients, so they reduce their activity and become dormant. Then another hormone called ethylene starts acting and causes the shedding of the tree because of the decrease in the physiological activity of the growth hormone called auxin. We can see then that it is a complex process involving three plant hormones but triggered by ABA due to environmental factors, light, temperature or lack of water in both dry and cold ecosystems. However, it is common in ecosystems like the one in the city of Lima.

But all plants do not become dormant. It depends on genetics of the species. Therefore, we can classify the plants in those that shed leaves and those that do not (see Table 3.3). Based on leaf persistence, trees can be classified as follows.

Deciduous

Trees that shed their leaves one season a year due to change in the weather or in the amount of water they get. These trees are common in temperate or dry tropical climates with long periods

Table 3.3 Evergreen and deciduous trees used in gardening in Lima, own authorship

Evergreens and deciduous trees used in gardening in lima	
Deciduous	Evergreens
Poinciana: *Delonix regia*, native to dry ecosystems in Madagascar, naturalized in central and South America	Rosewood: *Tipuana tipu*, native to South America
Kapok: *Ceiba trichistandra*, native to dry ecosystems in the north of Peru	Rubber: *Ficus elástica*, native to tropical ecosystems
Fern tree: *Jacaranda mimosifolia*, native to tropical and subtropical ecosystems	Hawaian Molle: *Schinus terbinthifolius*
Apple: *Malus domestica*, native to Asia Minor	Palms, all species, most are native to Asia
Pear tree: *Pyrus communis*, native to Asia Minor	Cypress: *Cupressus semprevirens*, Native to the mediterrean
Pecan tree: *Carya illinoinensis*, native to south east US	Chilean pine: *Araucaria araucana*, native to South Chile
Kapok: *Ceiba trichistandra*, native to dry forests in the north of Peru	Mountain Eucalyptus: *Eucalyptus globolus*, Coastal Eucalyptus: *Eucalyptus camadulensis*, native to Australia
African Tulip: *Spathodea campanulata*, introduced species	Yellow Trumpetshrub: *Tecoma stans*, native to the lower mountains in Peru
Palo verde: *Parkinsonia aculeata*, semicaducifolia native to the dry forest	Mountain Molle: *Shinus molle*, native to Riverside forest ecosystems

of drought. Most deciduous trees used in design are not native to South America. Deciduous trees in climatic conditions like Lima's do not shed all their leaves in a certain season because there are no marked differences between seasons. This causes a poor development in the tree since it cannot recover during dormancy from the intense period of growth common in this specie. Deciduous trees outside their place of origin may go without a period of dormancy which makes them tired and causes them to lose strength quickly. They require a period of rest to recover and start growing vigorously again when their period of physiological activity resumes and they start producing clusters of flowers in Spring. In Lima's climate, dormancy could be achieved by not watering one season a year. If watering is not suspended, trees will be weak and flowers small and sparse throughout the year. The most common tree is the apple tree (*Malus domestica*) that under the climatic conditions in Lima produces small apples of poor quality. Another tree that reacts in a similar way is the poinciana (*Delonix regia*).

The benefit of deciduous trees when properly managed is that they give shade all summer and light all winter. Furthermore, they help parks stay clean and do not attract spiders and insects. The drawback is that in winter, trees will have no leaves giving the environment a sad and cold look.

Evergreens

They are trees that do not go dormant. Evergreens are not affected by changes in the environment since they do not have strong and marked hormonal variations. They change their leaves slowly throughout the year, so they always have new foliage and do not stop doing photosynthesis. Most evergreens are from tropical and subtropical humid areas. When these trees are planted in cold places, they soon perish due to hydric stress caused by cold winter draught since water freezes and cannot be used for watering. This also happens in dry climate where there are draughts if plants are not watered. This is the case of most ecosystems in Lima. However, when trees are native of the area, they will be able to survive.

3.2.1.1 Tree Pruning

The purpose of pruning is modifying or preserving the form of plants. It can improve production and facilitate harvesting, eliminating pests or preserving crown form and ensuring tree balance. Therefore, it is an important agronomic practice. Pruning has been used for a long time as an agricultural technic for fruit production.

In the year 79 a.C. Pliny the Elder, defined the term topiary *(opus topiarium)* to refer to ornamental pruning (Kluckert 2000) which is a practice used now more than ever. Francesco Colonna in his book *Hypnerotomachia Poliphili* in 1499, suggests pruning plants by modifying their natural form into different decorative shapes (Kluckert 2000). Topiaries were used in Hampton Court Palace by King Henry XVIII who enjoyed palatial gardens. However, this technique had detractors, such as Francis Bacon who in 1625 published his book *On the garden's* in which he opposes to the use of topiaries because he considers they go against the natural form of plants (Kluckert 2000).

Concerning ornamental pruning, landscapists have divergent ideas about pruning in gardens. Classical landscapists use pruning to give order and organize the garden, including topiaries and modifying the natural form of the plant. On the other hand, natural landscaping defends the use of the natural forms of plants. Pruning has other important roles in park management. The plants are not in their place of origin and most species used in the city have gone through a process of genetic selection and after leaving the plant nursery, they are modified to prevent them from growing too much which could cause problems in the city. Park trees need maintenance; old branches should be pruned, and trees should be balanced to prevent them from falling.

There are different types of pruning based on purpose. Thus, we can list: formation pruning, maintenance pruning, production pruning, cleaning pruning. The purpose of each type is as follows:

- **Formation pruning**, consists of eliminating the tree's terminal bud. This pruning should be done in the plant nursery, before the tree is one year old. This can also be done to the root of a deciduous tree. The terminal bud of the root can be eliminated to stop radicular expansion and prevent the tree from growing. Formation pruning should be done before Spring. In Lima, it should be done in August, so that when spring arrives, wounds close quickly. When a tree is subject to formation pruning, it can be up to 30% smaller. That is why, this type of pruning is frequently used in almost all trees suitable for urban gardening. When the terminal bud is eliminated from the plant, its hormonal concentration is also modified. The plant hormone auxin concentrates in the terminal bud and inhibits growth of the lateral buds, so growth is centred in the terminal bud which quickly directs it upwards. When the apex is eliminated, side branches are encouraged to grow. Normally, young plants have more gibberellines, hormone that causes cell division and lengthening, concentrating auxin in the terminal bud of both the crown the root so trees grow vertically quickly in order to compete for the light with other trees. When they reach adulthood, levels of auxin increase and the tree stops growing upwards and begins growing vertically, thus, the crown becomes bigger. In this way, established adult plants, mark their space and prevent other plants from growing under their crown. With formative pruning, growth is modified by encouraging the formation of lateral branches.
- **Maintenance pruning** intends to preserve the form of the tree once it reaches adulthood and prevent it from growing too much. In this sense, it is like a haircut, Maintenance pruning should be done once a year in August. It is gentle and intended only for the elimination of damaged or bent branches or those that are not part of the form of the tree as well as for old leaves that could house insects or pests. This will enable the light to penetrate the inside of the tree in spring so that there are no barren parts where no light gets in. Deciduous trees can be pruned more drastically because they do not have leaves in August. However, evergreens should be given gentle and

frequent maintenance pruning to not affect tree growth which depends on leaves.

- **Production pruning** is intended for good harvesting in the case of fruit trees or for good blooming in the case of ornamental trees. To produce flowers and fruit, a tree requires nutrients. If it does not get proper nourishment, it will get weak flowers or fruit that will fall easily. It is estimated that each flower requires at least three leaves for proper development of flowers and fruit. However, this could change depending on the species. Light is also an important factor for production. When a tree does not get enough light within, the part with no light will be unproductive. Branches will elongate in search of light, producing fruit and flowers only at the end sections whereas the centres will be barren. No more than one third of a tree can be pruned or it could die. It is important to distinguish deciduous trees from evergreens since deciduous trees bloom from adult branches of one or two years old, depending on the species. Therefore, if all terminal branches are pruned, we will not have flowers so only one out of two terminal branches could be pruned. For fruit production, the age of the branches should be consulted, according to the species.
- Cleaning pruning seeks to eliminate and prevent pests. This is done by eliminating the parts of the plant that form compact shrubs and prevent the sun from coming in. Old leaves and branches should be eliminated to. Cleaning pruning should be moderate. If pruning is excessive, the tree will not have time to recover and will weaken due to lack of nutrients since with few leaves it will not be able to do proper photosynthesis and can be affected by pest. It is advisable to fertilize the tree after pruning with organic fertiliser placed in a circle 5–10 cm deep around the tree.

3.2.1.2 Use of Trees in Urban Landscapes

Trees have different uses in urban landscape. They can be used as **windbreakers**, but they should grow straight to be effective, so, especially if the wind is strong, they will need support poles for the trunks to prevent them from bending (see image 3.5). A key point is to select a tall species with a columnar crown and a straight trunk. Such as the Black Poplar (*Populus nigra*), the Humboldt Willow (*Salix Humboldtiana*), the Mountain Eucalyptus (*Eucalyptus globulus*), the Coastal Eucalyptus (*Eucalyptus camaldulensis*) or the Casuarina tree (*Casuarina equisetifolia*). Smaller trees could also be used such as the yellow trumpetshrub (Huaranhuay) (*Tecoma stans*).

In several cases, since trees grow slowly, they can be planted with fast growing shrubs which could be used as defensive fences. Ideally, these shrubs should not be dense or impenetrable, but translucent, enabling the wind to go through slowly without breaking or toppling the tree. Shrubs that could be used are the huaranguillo (*Acasia horrida*), the chinese hibiscus (*Hibiscus rosa sinencis*), retama broom tree or genus (*Retama sphaerocarpa*), Buxus (*Buxus siempervivens*), Brazil raintree (*Brunselfia pauciflora*); all these plants can adapt well to the dry conditions of Lima.

Wind-breaking trees will also help reduce dust and dirt in the garden. The trees cannot be too dense, especially if they are combined with shrubs because the wind will bounce and intensify on the side the wind is blowing forming wind swirls or increasing in an area 10 times bigger that the size of the tree.

Trees are also used along highways or wide avenue. In this case, it is very important to consider visibility. Along a highspeed highway, trees can block the headlights of cars coming in the opposite direction, making it uncomfortable to drive. To prevent this, it is advisable to combine trees with shrubs or use columnar trees and give them maintenance pruning to keep their branches low. Plants chosen should also grow easily in the dry environment of Lima without requiring much maintenance or watering. Consequently, the highway will not need to be blocked often. In addition to maintenance pruning, the trees will need formation pruning of the crown once a year in August. Trees suitable for highways along the coast are: trees with a low, dense crown like the mountain mole (*Shinus molle*), the mimosa (*Acasia macracantha*), the carob tree (*Prosopis pallida*), the cypress fence (*Cupresus leylandii o Cupressus macrocarpa)*, the Casuarina (*Casuarina equisetifolia*).

The trees in the city should allow visibility for both vehicles coming in the opposite direction and pedestrians. They should be placed at least 3 m away from the corner to enable safe turning. Their crown should be high, and their roots small and superficial so that they do not break sidewalks, pavements or sewage pipes. Therefore, when choosing a tree, we should consider the form of the crown, the speed of growth, the depth of the roots, and its resistance to pollution. The roots should be modelled in the plant nursery before it is one year old, preferably when it is 3 months old. To do so, the apical bud at the tip of the root is cut which causes the root to fork and develop sideways and stop growing downwards. Consequently, the tree's root depth will be one third smaller than that of a tree in its natural form. Evergreens are trees resistant to pollution whereas deciduous are not because shedding their leaves makes them more sensitive. On the other hand, the trees should be suitable for the ecosystems in Lima which have very little water but a lot of air moisture (see Table 3.4).

When planting a tree, it is very important to be aware that the tree will grow in the same place for decades. Therefore, it needs deep soil, properly drained and fertilised with plenty organic matter. Synthetic fertilisers are hazardous to the environment and if used excessively damage the groundwater. Soil should be prepared adding compost or manure to the bottom of the hole before planting the tree. Every year, before Spring, trees should be fertilised with organic matter like compost, manure or humus.

3.2.1.3 Ways to Prevent Damage Caused by Salts

In Lima trees can be affected by the abundant salt in the soil or in the marine breeze. During evaporation after watering trees, salt that is at a certain depth in the soil is transported to the surface, especially in summer. If we prevent evaporation by irrigating with the exact amount of water required by the plant, for example, using drip irrigation or isolating the upper layer with an insulating cover such as a thin mesh, we will prevent damage caused by salt in the subsoil. Soil in Lima can be of two types depending on its origin: alluvial, brought by the river in the areas that form the Lima valley, or saline original of the area. Alluvial soil is fertile and does not have salt except for that in the marine breeze whereas the soil that is original of the area has salt.

Marine breeze brings the salt form the sea onto the leaves burning them. In Central America where it rains intensely, the salt washes off without damaging the plan. Whereas in Lima, since there is very little rain, leaves are badly damaged. However, we can prevent this by washing the leaves maybe through drop irrigation or simply washing the plants with sprinklers twice a week. This cannot be done when the sun is blazing hot because leaves will burn.

Plants respond to salt in different ways. There are plants that have adapted to grow in saline soil. Others, on the other hand, are very sensitive and do not develop well in saline soils. Plants that have adapted to saline soil have thick leaves covered with wax, thus, they are not hurt by

Table 3.4 Sturdy trees suitable for the city of Lima, own authorship

Sturdy trees suitable for Lima city	
Trees that do not require much water	Trees resistant to pollution
Mountain Mole: *Shinus mole*	Carob: *Prosopis pallida*
Palo verde: *Parkinsonia aculeata*	Palo verde: *Parkinsonia aculeata*
Mimosa: *Acasia macracantha*	Ceiba: *Ceiba trichistandra*
Yellow Trumpetshrub: *Tecoma stans*	Palo verde: *Parkinsonia aculeata*
Rose Wood: *Tipuana tipu*	Weeping Willow: *Salix babilónica*

Table 3.5 Trees that tolerate soil salinity and marine breeze, own authorship

Trees that tolerate soil salinity and marine breeze	
Fig tree	Ficus carica
Fan palm	Washingtonia filifera
Casuarina	Casuarina equisetifolia
Carob tree	Prosopis pallida
Jerusalem Thorn	Parkinsonia aculeata
Chilean pine	Araucaria araucana
Myoporum	Mioporum laetum
Dracaena tree	Dracaena draco

marine breeze and they prevent salt from reaching the roots as well. Furthermore, these plants change leaves often to eliminate the salt. Trees selected for Lima should tolerate soil salinity and marine breeze (see Table 3.5).

3.2.2 Shrub

A shrub has several branches of similar width stemming from the base, whereas, a tree has a main trunk. This feature is determined by the plant's hormonal distribution defined by genetic factors that express themselves based on environmental conditions. Unlike shrubs, trees have apical dominance caused by the apical production of auxin in its terminal buds both in the root and crown. Therefore, shrub branches grow sideways from an early age. The lack of apical dominance is important to consider when pruning. Shrubs can be pruned more easily and drastically without causing harm since lateral branches will sprout again quickly.

Shrubs can determine vertical and horizontal forms in a garden depending on location, pruning and whether they are placed together or individually, as well as on their natural form. Shrubs are the wall of our design. They are also used to divide spaces, form paths or give colour by forming groves. When shrubs are used to form paths or groves, the garden will appear horizontal. However, if they are planted in small groups or individually, it will appear vertical.

Shrubs can grow in two ways: compact growth with leaves from the base or umbrella growth with no leaves coming from the base. This depends on the concentration of the vegetable hormone auxin, that prevents leaves from falling or abscisine hormone that promotes it. The two types of plants can be combined, or the empty area of the umbrella shaped shrubs could be covered with mid-size plants or flowers.

Shrub management is not simple; when it is not done properly, it can be boring; for example, if we use leaf shrubs placed against a long wall and all cut the same size. In this case it is advisable to plant groups of shrubs of different tones and varieties but similar texture and forms. On the other hand, if shrubs of different varieties are placed along the same wall it will look disorganized, even worse if we use shrubs with flowers of different varieties and colours. The form and texture of shrub leaves is very important. They should combine with the architecture if we want to enhance it. Big leaf shrubs should be used for rustic architecture, whereas slick-leaved shrubs will make a house look tidier and better kept and combine well with smooth and uniform material. However, if we want to draw attention and create a contrasting effect with the architecture, we should use plants with discordant texture which will give a sensation of bewilderment, often used in pieces of art to create divergence between elements.

It is common to use shrubs to outline paths or areas, but it is only advisable in a big garden or the garden could look even smaller than it is. Shrubs could be a perfect choice to outline a football field or separate the playground. For each case, we should consider the height of the shrub as well as transparency. For example, in case of a playground shrubs should not be dense so that it is still visible on the other side. A curtain of transparent shrubs would make the space look bigger. Compact shrubs are perfect to make mazes in a playground or to design sculptures. For example, the fence cypress (*Cupresus leylandii o Cupressus macrocarpa*), the dwarf granade (*Punica granatum* var. nana), or the benjamine ficus (*Ficus benjamina*).

When planting shrubs, the fourth dimension should be considered. Therefore, size is crucial. Since shrubs can be between 50 cm and 3 m tall, the space where they are going to be planted should be examined, considering that the roots will be at least one third the size of the plant. Another factor to be considered is time; shrubs live up to 30 years so soil should be carefully prepared. Shrubs should not be planted next to a wall where the soil is acid since it has remains of construction material and cement. Soil should be prepared up to a depth and width of 50 cm.

Shrubs can be selected based on the colour and texture of their leaves or the colour and fragrance of the flowers as well as the combination of colours of leaves and flowers. The colour of the leaves might change slightly depending on sun exposure and amount. For example, when shrubs have a purple hue and they are placed in the shade, they lose their purple hue and become dark green. Likewise, yellow or red crotons (*Codiaeum variegatum*) can lose their colour if they are placed under too much shade.

It is important to consider that shrubs are not prioritized in park design policies in cities. This is the case for both Warsaw (Swoczyna et al. 2017), and Lima. In the case of Lima, municipalities as well as neighbours prefer not to plant shrubs because of vandalism and lack of security. However, shrubs are planted in many parks of Lima around certain areas of grass to prevent people from stepping on it, rendering these areas impractical for public use since they cannot be accessed or viewed.

Most shrubs are planted for their colours, like Crotons (*Codiaeum variegatum*) or Breynia (*Breynia nivosa*), but there are shrubs that bloom all year around, for example the Hibiscus (*Hibiscus rosa-sinensis*) or others with flamboyant fruit like the Pomegranate (*Punica granatum*) (see Table 3.6).

Table 3.6 Shrub classification based on use, own authorship

Shrubs by colour		
Leaf colour	Common name	Scientific name
Yellow and green	Lemon croton	*Codiaeum variegatum*
Red, lilac and yellow	Croton	*Codiaeum variegatum*
Green with yellow edge	Acalypha	*Acalypha wilkesiana*
Red or white with green	Breynia	*Breynia nivosa*
Green or yellow with green	Ficus benjamina	*Ficus benjamina*
Colour of Flowers		
Light blue	Plumbago	*Plumbago auriculata*
Orange	Chinese lantern	*Abutilon pictum*
Red	Bottlebrush	*Callistemon citrinus*
Red, yellow, white or orange	Rose mallow	*Hibiscus rosa-sinensis*
Yellow	Spanish broom	*Spartium junceum*
Red and White	Nerium or oleander	*Nerium oleander*
Lucuma	Rose or dwarf laurel	*Nerium oleander*
White	Angel's trumpet	*Brugmancia arborea*
Red, red with lilac or white	Scarlet Fuchsia	*Fuchsia coccinea*
Modified red leaves, simple white flowers	Poinsettia	*Euphorbia pulcherrima*

3.2.2.1 Shrub Classification Based on Pest Tolerance

Any plant subject to an inhospitable environment is more sensitive to pests and disease. Especially if the soil is poor and watering sparse, there are high probabilities that the plant will be affected by pests. Due to inadequate conditions for growth, the plant tends to bloom and produce fruit in order to perpetuate the species depleting nutrients needed for growth and concentrating on the fruit. Therefore, it is important to improve the life conditions of the plant by fertilising it with organic fertiliser and watering it enough. However, if the plant is fertilised excessively, it will absorb it for its own growth, delaying blossoming or not blossoming at all.

Shrubs are pruned to control growth, to keep shrubs a certain size and to keep the area free from pest. Most shrubs are pruned frequently which alters their original size, However, this shortens their life and increases the risk of catching disease. Furthermore, it affects flowering since most shrubs have flowers at the end of new terminal branches; especially those used in parks in Lima. Light is essential in the development of pests. Too much light or too little can cause one; if we add infertile soil, the plant will most certainly be affected by pests. When a shrub is too dense, it will have barren areas with no light. These spaces house both spiders and insects, Therefore, one of the main objectives of pruning is eliminating compact parts that prevent the sun from going through. Shrubs that are pruned excessively weaken since they cannot recover fast enough due to lack of nutrients caused by its inability to do photosynthesis when it has only few leaves and consequently are easily attacked by pests.

When selecting shrubs, we should consider tolerance to pests and disease, light requirements and speed of growth. Some shrubs are very sensitive to pests. A common pest, very difficult to fight in warm cities like Lima is the Scale insect, which places itself at the fork of branches and roots, so it is difficult to eliminate. It is vital for plants to be properly nourished in order to resist pests and disease (see Table 3.7).

For areas in Lima close to the ocean it is important to consider tolerance to salt. Plants subject to marine breeze are also sensitive to pests and disease. It is important to water plants early in the day in order to eliminate salt from marine breeze accumulated the evening before. Salt prevents plants from absorbing water from the soil and burns leaves clogging their pores. In the Table 3.8 are listed the Shrubs resistant to marine breeze.

3.2.3 Vines

Botany defines vines as plants with indefinite lateral growth which means, that the growth of their apical bulb dominates lateral growth during their entire life. The main stem continuous growing after the plant's youth period is over, unlike trees that have a definite growth period. This is due to the hormonal distribution of the plant. In nature, vines climb up trees to reach the light and do not need to develop a trunk. In this

Table 3.7 Shrubs based on resistance to pest and disease, own authorship

Shrubs based on resistance to pest and disease	
Shrubs resistant to disease	Shrubs sensitive to pests and disease
Sleeping Hibiscus *(Malvaviscus arboreus)*	Chinese lantern (*Abutilon pictum*)
Rose Mallow (*Hibiscus rosa-sinensis*)	Breynia (*Breynia nivosa*)
Nerium or Olander (*Nerium oleander*)	Scarlet Fuchsia *(Fuchsia coccinea)*
Dracaena (*Dracaena sp.*)	Lantana (*Lantana cámara*) con exceso de sol
Myoporum (*Myoporum laetum*)	Golden Shrimp (*Pachystachis lutea)*
Buxus (*Buxus sempervivens*)	Marmalade (*Streptosolen jamesonii)*

Table 3.8 Shrubs resistant to marine breeze, own authorship

Shrubs resistant to marine breeze	
Common name	Scientific name
Seagrape	Coccoloba uvifera
Nerium or oleander	Nerium oleander
Angel's trumpet	Brugmancia arborea
Spanish Broom	Spartium junceum
Ngaio	Mioporum laetum
Sleeping Hibiscus	Malvaviscus arboreus
Rose Mallow	Hibiscus rosa-sinensis
Buxus	Buxus sempervivens
Fence Cypress	Cupressus macrocarpa

way, they win the battle towards the light. They grow quickly and when they reach the top they blossom. Therefore, most vines require light to blossom and they do so on new branches. They also have flamboyant colours to attract pollinators.

Vine stems can be herbaceous or woody which is easily distinguished; herbaceous vine stems are green. They have a short life span of two or three years and grow quickly. On the other hand, woody vine stems are long and vegetative. They live between 20 to 30 years, depending on the species and living conditions. Woody vines can be pruned to form shrubs like the Bougainvillea *(Bougainvillea spectabilis)*.

Some vines have rings, which are simply modified leaves, to help them climb trees and walls. Certain rings or adventitious roots release a sticky substance that help plants adhere to the wall, such as Ivy (*Hedera hélix*) or the swiss cheese plant (*Monstera deliciosa*). Others form carbonic acid by combining carbon dioxide with the water in the environment causing damage to walls, stones or sculptures.

Based on the way the vine grows they can be classified as follows:

- Scandent vines, with woody stems that enable them to be semi-erect without support until a certain height, such as the wild rose (*Rosa sp.*), the golden trumpet (*Allamanda cathartica*), the Bougainvillea *(Bougainvillea spectabilis) or the ylang-ylang* (*Cananga odorata*).
- Vines with fragile stem that twine around a support, such as the sweat pea (*Lathyrus odoratus*).
- Vines that climb up walls with their rings such as the Mexican coral vine (*Antigonon leptopus*).
- Vines that release a substance to adhere onto a wall such as ivy (*Hedera helix*).

3.2.3.1 Vine Pruning

Vines can be managed by pruning. In places located on the coast, such as Lima, vines should be pruned when winter is over at the end of August or at the start of September, when the plant is growing slowly and has few new leaves. In this way, the plant will lose fewer nutrients during pruning and will recover more quickly when it restarts growing in Spring. If pruning is done at the start of winter or in the middle of winter, it will heal more slowly and might show health problems, especially fungus. When pruning, it is important to consider that the plant should be in balance with its roots. If roots are still not fully developed, branches could make the plant topple if they are all directed to one side.

For formation pruning intended for a vine arbour or roof, all branches before the start of the arbour should be removed, choosing between three to five of the top secondary branches. The ideal plants for this purpose are the woody vines. The branches chosen should not be opposite to each other or form between both an angle of 180 degrees since they could break in the point of union with the main branch. The secondary branches should be distributed evenly to allow air and light to go through. If not, these branches will not be properly nourished and will be weak, without leaves, thus, more sensitive to disease and pests. Only woody vines can be subject to this type of pruning. Herbaceous vines like the sweat pea (*Lathyrus odoratus*) or the coral vine (*Antigonon leptopus*), cannot be used for an arbour. However, they can complement the design by being placed laterally if a smaller space is desired. Notwithstanding, because of their short life span, they will need to be replanted.

Formation pruning to cover a wall with a mesh vine requires pruning of the terminal part of the plant so that the secondary branches start dividing which marks an end to the plant's youth, stimulating the production of flowers and fruit. The lower the main branch is cut, the lower the secondary plants will begin to grow, hence, the better the mesh will cover the wall. This is the reason this type of pruning is done to cover complete walls. Herbaceous vines grow quickly and do not need pruning except for cleaning pruning, these form a beautiful mesh that covers the wall naturally but needs to be replanted every certain amount of time which can happen naturally when seeds sprout in the soil and replace adult plants.

When terminal branches are cut, it triggers new growth and renews the oldest buds in the bottom part of the plant which enables vigorous growth as well as eliminates pests and disease and makes cleaning easier. All cuts should be diagonal to prevent water stagnation. The section located 2 to 3 cms below the cut will die and give place to the branch's callous. For this reason, the cut should not be at the edge of the bulb but a couple cm above it to prevent a hole from forming in the stem where humidity could accumulate and attract fungus, insects and pests. Once the plant's structure is formed, pruning should be done annually to keep the secondary branches formed.

The best vines for arbours are grapevines, bougainvilleas *(Bougainvillea spectabilis)* and the yellow trumpet (*Allamanda cathartica*), among others.

Most times vines are planted in a garden to cover a wall quickly. They can also form an arbour, or pergola or they can be used to form a fence so that a wall cannot be climbed over. Before planting a vine, we should consider how it will grow, the colour of the flower, life span, time of blossoming and speed of growth. By doing this we will be able to choose where to plant the vine based on why we chose it. Vines have flamboyant and colourful flowers, but many do not bloom all year around. Therefore, to design a garden it is important to know when the vine blooms (see Table 3.9).

3.2.4 Green Beds

Garden beds are groups of plants placed together to enhance their form or colour. They can have from 3 to 7 plants depending on their size or they can even form larger groups of several tens. Garden beds come from the XIX century and were used in classical gardens. They were introduced by John Claudius Loudon (1783–1843) who suggests the use of plants of the same species to form what we know today as garden beds or flowerbeds resulting in big fields with groups of plants or flowers (Kluckert 2000: 396).

This concept originates from considering agriculture as part of a design. Poet William Shenstone (1714–1763) was one of the practitioners of this style who introduces aesthetic principles into a farming estate (*Ferme Ornée*) or ornamental farm where the countryside is seen as an idyllic space. Following this line, landscaping in France does not only incorporate spaces for farming but also areas for grazing, recreating the pastoral world. A renown garden that picks up this idea is Ermenonville where the garden described by Rousseau in his book *Nouvelle Heloise* is recreated as heaven on earth (Kluckert

Table 3.9 Vines based on time of blossoming, own authorship

Vines based on the time of blossoming	
Summer	
Common name	Scientific name
Coral Vine or mexican creeper	*Antigonon leptopus*
Honey suckle	*Lonicera periclymenum*
Bougainvillea	*Bougainvillea spectabilis*
Cress	*Tropaeolum majus*
Spring	
Golden Trumpet	*Allamanda cathartica*
Coral vine or Mexican creeper	*Antigonon leptopus*
White jasmine	*Jasminum officinale*
Autumn	
Golden Trumpet	*Allamanda cathartica*
Glory-bower	*Clerodendrum speciosum*
Bougainvillea	*Bougainvillea spectabilis*
Beach moonflower	*Ipomoea violácea*
Honeysuckle	*Lonicera periclymenum*
Flame vine	*Pyrostegia ígnea*
Winter	
Glory-bower	*Clerodendrum speciosum*
Flame vine	*Pyrostegia ígnea*
Bougainville	*Bougainvillea spectabilis*

2000: 441). Likewise, park Raubouillet where Luis XVI builds a *ferme ornée* for Marie Antoinette which we can describe as an idyllic farm landscaping, including a merino sheep farm and a dairy (Kluckert 2000: 443).

The use of farming in design has returned with designers such as Anna Scaravella, Fernando Caruncho, Peter Walker, Martha Schwartz, among others. These designers use geometry common in agriculture as well as farming plants (Kluckert 2000: 258–261). Nowadays, one of the problems in Europe is the disappearance of countryside landscape and the reconquest of the forest due to the abandonment of farms. Therefore, these designs represent a way of cultural landscaping, planting big garden beds in strips with the intention to preserve the cultural aesthetic landscape.

Garden beds can be used to decorate spaces, combining groups of colours, and giving the garden volume. Depending on the garden, garden beds can be formed by a paired number of species of similar size and shape species, as is the case of classical formal gardens, or an odd number of species of different size and form, such as the modern, informal gardens that avoid repetition of similar spaces. Most shrubs can be used to form garden beds and most season flowers can form colourful flowerbeds. It is important that in a garden bed all the plants can be given proper maintenance, especially in the centre. Therefore, a garden bed should not have a radius bigger that 90 cm. Plants in a flower bed can be woody our herbaceous, wild or cultivated (see Table 3.10).

Woody garden beds can be planted under a tree, but they will be difficult to care because they are perennials with relatively big roots that might compete with the superficial roots of the tree; however, they are a good alternative from a

Table 3.10 Garden beds according to their form of growth, own authorship

Garden beds according to their form of growth		
Woody garden beds	Herbaceus garden beds	Bulb garden beds
Hydrangea: *Hydrangea macrophylla*	Bernard's lily or spider plant: *Chlorophytum comosum*	Eucharis or amazon lily*: Eucharis x grandiflora*
Ixora*: Ixora coccinea*	Heliconia or False bird of paradise*: Heliconia sp.*	Amancay: *Ismene Amancaes*
Marmalade: *Streptosolen jamesonii*	Alteranthera: *Alternanthera reineckii*	Daffodil: *Narcissus pseudonarcissus*
Scarlet Fuchsia*: Fuchsia coccinea*	Gerbera daisy*: Gerbera jamesonii*	Nard: *Hippeastrum miniatum*
Elephant-ear*: Colocasia sp.*	Anthurium: *Anthurium andreanum*	Agapanthus or African lily: *Agapanthus africanus*
Golden Shrimp*: Pachystachys lutea*	*Red sage: Salvia splendens*	Iris: *Iris germánica*
Asparagus Fern: Asparagus densiflorus	Coleus*: Coleus blumei*	Tuberose Vara de San José: *Polianthes tuberosa*
Geranium: *Pelargonium hortorum*	Canna or Indian shot*: Canna indica*	Lily: *Lirium longuiflorum*
Lantana*: Lantana camara*	Wallflower: *Matthiola incana*	Tulip: *Tulipa gesneriana*

landscaping point of view. They require plants that need shade and little light such as ferns, ivy (*Pelargonium petate*), the swiss cheese plant *(Monstera deliciosa)*, depending on the design of the garden. Plants in the garden bed need enough nutrients both before they are planted and while they are growing. They need proper fertilizing with compost or humus at least once a year so that they do not begin a competition with the trees. Woody garden beds should be given a cleaning pruning. Other types of pruning vary from species to species. For example, geraniums (*Pelargonium hortorum)* should be pruned drastically but the lantana (*Lantana camara*) only needs pruning of its terminal section based on the height desired for the plant.

Since annual flowers are so flamboyant, it is common to choose them for flower beds, but they are difficult and costly to keep, unlike perennials that are a better choice. Pruning of herbaceous garden beds is difficult because they have a short life span. Therefore, it is more convenient to replant them annually than to prune them since pruning deforms the plant's original structure.

Bulb plants can be used for flower beds under trees. When they are under deciduous trees, they get light in winter and spring and give a little colour before trees get their leaves back.

Nards (*Hippeastrum miniatum*), Amancaes *(Ismene amancaes)* or the African lily (*Agapanthus africanus*) can be used.

3.2.5 Plant Coverage

Plant coverage is formed by a group of small plants that cover big spaces. Unlike ground covers, plant coverage does not endure high transit, although they are often referred to as ground covers. Many of them are blooming flowers and have a short vegetative period. Others can grow indefinitely and can stay in the same place for several years with a good thinning pruning. Some small shrubs can be used to form both garden beds and plant coverage depending on the number of plants used. Plant coverages use more space (see Table 3.11).

From the point of view of design, plant coverage form patches that give colour to the background of the design. They can be used in modern gardens as well as in minimalistic gardens. Nowadays, due to lack of space in house

Table 3.11 Classification of plant cover in Lima's parks, own authorship

Classification of plant cover in Lima's parks		
Semi-perennial	Annuals	Resistant salt
Crysanthemum (Chrysanthemum morifolium)	Wallflower (Mathiola incana)	Trailing ice plant (Mesembryanthemum spectabile)
Geranium (Pelargonium hortorum)	Marigold (Tagetes patula)	Alternanthera (Alternanthera reineckii)
Coleus (Coleus blumei)	Balsamine flowers (Impatiens balsamina)	St Bernard's lily Chlorophytum comosum
Cuphea (Cuphea hyssopiolia	Marigold (Tagetes patula)	Cuphea: Cuphea hyssopiolia
Alternanthera (Alternanthera reineckii)	Red sage (Salvia splendens)	Alternanthera (Alternanthera reineckii)
Yellow Lantana (Lantana camara)	Coloured cloves (Oxalis sp.)	Wondering jew (Zebrinia pendula)

gardens of big cities, ample gardens are designed including big visuals that enable the use of plant coverages. Like garden beds, they also need adequate maintenance to prevent pests and the shade that is formed beneath old branches, both in the case of garden beds and plant coverages. That is why it is essential to perform a cleaning pruning.

Cleaning pruning of coverage plants implies removal of old leaves and branches, especially those that gather below the branches to prevent pests and disease. In Lima the most frequent pests caused by plant density are screwworms and fungi.

3.2.6 Plant Coverage

Flowers are grown for their colours and fragrance. Flowers play an important role in industrialized cities, compensating people for a hectic city life away from nature. They give people another vision of space which is especially important for low income citizens and for those who put up with a lot of stress (Kluckert 2000: 397).

Flowers can be planted in flowerpots on windowsills, at the edge of gardens or in parks, under trees or umbrella shrubs and they can also be used as coverage. If they are used for edges, flowerpots on windowsills or flowerbeds, they should not occupy more than 90 cm wide, or maintenance could be difficult, especially for the part of the plant that is hidden.

Flowers are sensitive to fungus so the soil used should be slightly sandy. Furthermore, they grow fast, and their roots should be able to expand in the soil and grow quickly. We should not use moss because it has an acid PH that damages the plant and flowers and retains too much humidity. Furthermore, the nutrients in the moss are not used by the plant because moss has a prolonged time of decomposition.

Flowers come in three forms: the form of a spike like the wallflower (*Mathiola incana*) or the snapdragons (Antirrhinum majus); the most common round form like the marigold (*Tagetes patula*), the crysanthemum (*Chrysanthemum morifolium*) and the intermediate form like the iris (*Iris germanica*) (Muñoz 1979). Form should be considered in garden design. Round and spike flowers are used in classical gardens whereas intermediate flowers are used in modern designs, but this can vary depending on the design.

Based on their life span, flowers can be classified as annuals, bi-annuals and semi-perennials (see Table 3.12). Bi-annual plants can blossom twice or three times in their lifetime. The plants with semi-perennial flowers can live about 5 years. They reproduce from stolons so they can be found in the same place for decades. Many other plants give flowers. However, they are not considered flower plants because their periodicity is another and they are not planted with the

Table 3.12 Flower plants according to their lifecycle, own authorship

Flower plants according to their lifecycle		
Annual flowers	Semi perennials flowers	Bulb flowers
Zinnia (*Zinnia elegans*)	Petunia (*Petunia x hybrida*)	Eucaris or amazon lily (*Eucharis x grandiflora*)
Snapdragon (*Antirrhinum majus*)	Pansies (*Viola x wittrockiana*)	Iris (*Iris germánica*)
Sage (*Salvia splendens*)	Begonia (*Begonia semperflorens*)	Nard (*Hippeastrum miniatum*)

purpose of giving flowers; this is the case of many trees. Annual plants need to be changed once their lifecycle has ended. Therefore, a garden should only have a small space for them, or maintenance cost would be too high. To prevent empty spaces in the garden, normally flowers are first planted in a seedling nursery and are transplanted when they are 3 months old, just before flowering.

When a garden is designed, and the time for planting is chosen, it is important to consider when flowers will blossom. If they are seasonal, planting can be planned in such a way that there could be flowers in the same place all year long. Furthermore, the garden's focal point can be changed based on the colours of the flowers. If flowers are also wanted for vases, flowers such as snapdragons (*Antirrhinum majus*), Crysanthemum (*Chrysanthemum morifolium*), daisies (*Leucanthemum vulgare*), begonias (*Begonia semperflorens*), carnations (*Dianthus caryophyllus*) or wallflowers (*Mathiola incana*) could be planted (see Table 3.13).

Due to Lima's climate, with no drastic changes of seasons, many plants with a long vegetative period, do not have a definite blossoming period. However, excessive humidity in winter and excessive heat in summer do not help cultivation. Flower plants are attacked by insects, especially in the first stage of growth so they need a lot of care.

Because flower plants have herbaceous stems, their most common pests are aphids and fungi. Aphids are controlled with yellow traps which are panels covered in yellow plastic or fabric, coated with vegetable oil. Aphids are attracted by yellow and get stuck in these traps easily. Fungi is controlled by calculating the distance between the plant and irrigation. It is also important to remove old leaves and flowers before they are attacked by fungi. Flowers should be watered frequently but slightly. In Lima plants are especially sensitive to fungus in winter because atmospheric humidity is very high. That is why it is common to plant flowers in spring and autumn.

3.2.6.1 Flower Plants That Form Bulbs

Bulb plants come from cold or dry ecosystems. They stay underground most part of the year as bulbs, so we often think they are dead. That is why they should be planted in a space that can stay empty with the bulbs underground. This space could be concealed by a cover ground harmless to the bulb such as grass or Swedish ivy (*Plectrathus verticillatus*). After this period of rest, in Spring, the plant will blossom forming new bulbs for new plants. Most bulbs bloom in Spring, in August.

After blossoming, the bulb sub-divides. That is why bulbs should be separated every two or three years so that bulbs have enough space and flowers do not weaken. Bulbs require drained and sandy soil to prevent them from rotting. Once a year, after blossoming, it is convenient to dry them in a dry place wrapped in newspaper for one month. This will replace dormancy distinctive of cold or dry ecosystems. Once planted, they should be fertilised with worm castings and after a year they will bloom again.

So that a plant can bloom it needs to complete its vegetative period and receive a certain number of hours of light. This is controlled genetically

Table 3.13 Blossoming time for some flowers in Lima, own authorship

Blossoming time for some flowers in Lima		
Spring–Summer	Autumn–Winter	All year around
Pansies (*Petunia x hybrida*)	Phlox (*Phlox drummondii*)	Geranium (*Pelargonium hortorum*)
Wallflower (*Mathiola incana*)	Eucaris or amazon lily (*Eucharis x grandiflora*)	Iv (*Pelyargonium petate*)
Snapdragon (*Antirrhinum majus*)	Scarlet Fuchsia (*Fuchsia coccinea*)	Periwinkle (*Catharanthus roseus*)

and determined by the synthesis of a protein called phytochrome that acts under red light (daylight) and inhibits production under distant red light (nocturnal light). Some plants require longer days to bloom, whereas others only bloom when the days are short; that is why all plants do not bloom in the same season (see Table 3.13). Most flowers bloom in summer or spring. However, this varies depending on their origin and how they adapt to the environment. The amount of light they require depends on their origin; it is an ecological strategy of survival to prevent germination at moments when there is not enough heat or water, as it may apply, preventing the plant from dying.

3.2.7 Edges

As mentioned before, edges are to garden design what baseboards are to architecture; they divide a space and make it appear smaller. That is why, before designing edges, we should determine if we need a smaller looking divided space, or if we want to keep the space as an ample unit. Many time edges are used for practical purposes to prevent water from flooding. However, this can also be done by controlling the amount of water used in irrigation, thus, saving water which is especially important in Lima where water is scarce.

Most edges use semi-perennial, small plants which are easy to care for. Among the most popular plants we can mention the Alternanthera (*Alternanthera reineckii*) or Nerium oleander (*Nerium oleander*). Small flowers are often used when colourful edges are desired such as zinnias (Zinnia elegans), pansies (*Petunia x hybrida*), begonias (*Begonia semperflorens*), or marigolds (*Tagetes patula*) since their flowers are colourful.

Appendix 1 Gives More Details About This Classification and the Use of All the Groups Described in This Chapter.

3.3 Function of Each Part of the Plant

Every living being is born, grows, develops and finally dies. Plants need to go through these stages to complete their lifecycle. When we refer to growth, we only refer to the plant's increase of size, when we refer to development, we refer to hormonal changes produced by environmental factors. The first stage of the plant's growth is the vegetative period in which the plant develops its structure, leaves and roots that will serve to nourish and support the flowers and seeds. The next stage of development is the generative period in which the plant produces floral bulbs which ends when the plant begins blossoming, followed by the productive period in which the flower open and produce fruit and seeds. Finally, the period of senescence begins and then the plant dies. Plants with a short vegetative period bloom only once and then die. On the other hand, perennial and semi-perennials bloom every year repeating the different stages of development during all its lifetime. Every organ in the plant has a specific function so that a plant can produce a seed and perpetuate the species.

A seed is a miniature plant. It is already formed and has all its organs so that when it sprouts the seedling can immediately feed itself and survive. The parts of a seed are the cotyledons that could be one or two if the plant is a monocot or dicot respectively, the stolon and the radicle. The cotyledons are the seedling's reserve of nutrients and nourish it in its first stage of development when it cannot do photosynthesis. Once the chloroplasts start working, the leaf will produce nutrients and the root will enable the entrance of nutrients in the soil. Depending on the species in relation to the climate, the plant could activate first its aerial part, stem and leaves or root.

The leaf is the part of the plant where photosynthesis occurs by means of the chloroplasts in the leaf cells. The leaf regulates the intake of carbon dioxide and the release of oxygen and water through the stomates. When the stomates open, some water is released and enables the plant's evapotranspiration. This is essential in order to regulate the plants temperature since the water that is eliminated through evaporation allows an intake of fresh water from the root. That is why, in the ecosystems in the tropical rainforest the plants have a higher number of stomates than plants in cold ecosystems. This is the same process that regulates the intake of carbon dioxide and the release of oxygen during photosynthesis.

Leaves can also have specific functions. Such is the case of bulb leaves called catafylls. These leaves store nutrients for when the plants endure the cold in winter or the drought in the desert and enable them to withstand adversity. Other modified leaves such as scales, thorns or sepals protect the bud from the heat or the cold, the animals or the flowers respectively.

The stems and branches of a plant support the leaves, flowers and fruit. Nutrients circulate through the vascular bundles of the stem. There are two type of vascular bundles; phloem and xylem. Nutrients absorbed by the roots are transported along the xylem from the roots up to the other parts of the plant. Nutrients produced by photosynthesis called photosynthates are transported along the phloem.

Stems that stay green during the life of the plant, do not only support the plant but also do photosynthesis. They are especially important in places where leaves have modified into thorns to protect the plant form heat and prevent evaporation. In some cases, in cold climate, where plants reserve nutrients for when it freezes, some stems have transformed to store nutrients beneath the ground growing horizontally. This is the case of the potato, that has developed subterraneous stems that grow horizontally with branches that store nutrients for the cold season.

The root gives support to the plant so that it can stay straight and fixed to the ground. The shape of the root is that of an inverted glass. It is formed by the primary roots, secondary roots and root hairs.

The primary root grows upwards and determines the depth of the root. If the tip of the primary root is cut, the root forks and secondary roots develop on both sides of the primary root. This affects the entire plant and causes the growth of the plant's crown to reduce to one third the normal size of the species. The root hairs which can be found all over the roots, absorb the elements from the soil. Nutrients in the soil go through a strict process of selection by means of ion exchange in the plant. This means that the plant exchanges hydrogen ($H+$) and hydroxyl ($OH-$) ions, produced with the separation of the water molecule when it enters the leaf stomate; this exchange enables the intake of different nutrients to the plant. This process is vital to regulate the plant's internal PH balance and prevent an excess of ions. Once there is an intake of elements, they are transported along the xylem to the rest of the plant.

Once the plant reaches its productive stage, it gives floral buds that will later become flowers. This stage is activated by hormonal changes typical of genetics. It occurs on its own or dependant on the climate. When the plant is climate dependant, hormones may not activate if the climate is not adequate. This often happens in warm places like Lima with ornamental plants that come from cold countries. This is the reason tulips cannot be grown in Lima, but they can be grown in Cusco that has a colder climate.

The primary role of **the flower** is plant reproduction. The flower organs are the stamen, which is the pollen-containing organ and the pistil which is the female organ of the flower. Both are protected by the colourful corolla and the calix, commonly green. All the flower components are modified leaves. Most ornamental plants have complete hermaphrodite flowers and most on the same plant (monoecious plants), such as daisies (*Leucanthemum vulgare*), orchids, or wallflowers (*Mathiola incana*). But not all flowers are complete. There are flowers with female organs on one plant and male organs on a different plant (dioecious plants). All pines are dioecious as well as papayas (*Carica papaya*). A female plant will not give fruit if there are no male plants nearby.

The fruit protects the seeds of future plants. It is comprised by the exocarp or external hard layer which protects the seed; the mesocarp or middle layer that nourishes the seed before it can act on its own; the endocarp which is a thin layer that protects the seed directly and the seed itself which will become the new plant. Finally, the seed is protected by teguments or protecting wall. Some fruit is simple with only one seed, others are complex with multiple seeds. Some plants like pine do not bear fruit so they do not have a source of nourishment when they germinate.

The fruit ensures the plant's subsistence before and after germination. That is why many fruits have developed chemical and physical germination blockers. For example, many seeds are protected by fruit which will not allow germination without enough hours of heat or water that will eliminate other chemical germination blockers. Reserves in the fruit will ensure food for the first days after germination until the plant can do photosynthesis. Others like the orchid seeds germinate before they fall to the ground, since they carry out a symbiotic process with fungi ensuring nutrition.

3.4 Environmental Factors

A plant requires five environmental factors to grow and develop: nutrients, water, light, temperature and air. Only when these factors are in balance with the metabolism of each plant will they be able to grow and develop healthily.

Plants have a complex physiology with a variety of changes determined by plant hormones. Each stage of plant development has different hormonal characteristics. The type and amount of each plant hormone determines development and changes. The plant hormones determined genetically are directly influenced by the environment. For a plant to go from one stage to another, it needs an external stimulus such as temperature, light, water or others. In this way the hormones are activated to enable the development from the germinative stage to the productive one. The level of sensitivity to environmental factors depends on the genetic of each variety of plants influenced by the ecosystem of origin.

3.4.1 Nutrients

The plant requires 16 nutrients essential for its survival; most are absorbed by the root hairs and dissolved as salts by means of ionic exchange. For ionic exchange the cell exchanges hydrogen ($H+$) and hydroxyl ($OH-$) produced with the separation of the water molecule. In the plant, nutrients are transported by the xylem to the other organs. Along the plant xylem, nutrients move from an area of more concentration to an area of less concentration, dissolved in the xylem's water. The xylem is formed by lignified dead cells; the lignified walls form tracheids that prevent the column of water in the xylems from returning to the root and escaping from the plant due to gravity. Thus, the xylem tracheas protect the cell from draught, since due to excessive inverted pressure water in the xylem tends to come out from the cell and the plant. Only when the plant evaporates water by means of the stomates, will it be able to absorb more nutrients. Then, we can say that the absorption of nutrients is directly related to the opening and closing of stomates which in turn depends on external temperature.

Elements in the air such as carbon, nitrogen, hydrogen, oxygen and sulphur are absorbed by the leaves which use carbon, oxygen and hydrogen to produce sugars by photosynthesis.

These sugars form in cell chloroplasts and from there, they travel to the rest of the plant. The nutrients transported by xylems are distributed to each one of the plant cells. Then, there is a new intake of nutrients into the cell through ionic exchange ensuring a balanced PH. Depending on the plant's degree of adaptation to extreme conditions this can change, so the capacity of the plant to adapt to these conditions will enable its survival.

The nutrients can be classified, based on the plant's requirements, in macronutrients and micronutrients. The macronutrients absorbed by the roots are nitrogen, phosphorus, potassium, sulphur, magnesium and calcium. The ones absorbed by the leaves are carbon, oxygen and hydrogen. These nutrients are needed in large amounts. The micronutrients, required in very small amounts are iron, magnesium, manganese, boron, copper, zinc, chloride, molybdenum and nickel, which are also absorbed by the roots.

Nitrogen prevails in the vegetative organs of the plant. Therefore, it is needed in larger quantities during the first stage of plant growth or vegetative stage. Plants cultivated for the beauty of their leaves should be fertilised with enough nitrogen. However, too much nitrogen can delay blooming. On the other hand, not enough nitrogen will make the plant spindly and with very small leaves. The symptoms of a deficiency are easily noted by the light-yellow colour of the adult leaves.

Phosphorus plays a vital role in the absorption of nutrients, in plant reproduction and in radicular development. Phosphorus is used in the plant's respiration process, since it is a component of ATP (adenosine triphosphate) which while transforming into ADP (adenosine diphosphate) releases energy used by the plant in vital processes. It is also present in nucleic acids that form the nucleus and in the phospholipids that form and protect the cellular membrane. Plants with a phosphorus deficiency have poorly developed roots and they are weak when attacked by pests. This deficiency can be distinguished if leaves turn purple, especially adult leaves.

Potassium is also essential for the development of a plant. It has a crucial role in the regulation of the opening and closing of the stomates. That is why a plant is resistant to draught and frost, since it indirectly regulates plant's temperature and the flow of nutrients. When stomates do not open, water does not enter the plant and nutrients will not flow. Leaves of plants with potassium deficiency have silvery brown, burnt and brittle tips.

Magnesium deficiency is frequent in plants. There is an antagonism between magnesium and calcium in the soil. Magnesium is a component of chlorophyll, and several enzymes. Chlorophyll is composed by iron and magnesium. Therefore, plants with magnesium deficiency show interveinal chlorosis or a yellowish colour between the veins of the leaves.

Plants usually do not have a deficiency of other macro elements absorbed by leaves nor do they have a deficiency of nutrients required in small quantity. However, in parks and gardens in the city, plants do not have access to agrarian soil like in the countryside, but they are in soil with pollutants so they might have a deficiency of micronutrients.

In general, to prevent a lack of micronutrients it is advisable to add organic material at least once a year since it contains all micronutrient.

The balance of soil and fertility is kept when all the elements in the soil are recycled in order to ensure their permanence for a long time. This does not occur naturally in an urban ecosystem. Therefore, we should add elements like moss, compost or humus to keep the soil fertile throughout time. It would be ideal for the municipalities of Lima to manage the production of compost; some already do it and it also could be done in house gardens and bio orchards on rooftops.

If one of the 17 elements is missing, or if there is too much of any of them, the plant will show deficiencies not only of the missing element but also of other elements that might be blocked by the elements in excess. Adding too much of an external element to the soil, can break the balance or produce a deficiency of the other elements and change the soil PH as well as compete to enter the plant when it has the same charge, thus, it could destroy the balance of the ecosystem. In brief, there must be a balance among the elements in the soil. Adding any element affects the others. When there is deficiency of one of the elements in the soil, this element could be there but blocked by another element, forming chemical substances in the ground, thus, unable to enter the plant.

That is why we require the minimum necessary of each one of the elements so that others are not missing. This principle is called the **law of minimum** and it applies both to the soil and to all the elements that form the ecosystem (see Fig. 3.7).

In urban ecosystems plants also get external nutrients through fertilisers added to the soil. These could be organic or inorganic, natural or produced in a laboratory. Plants grown in the city usually have a limited amount of nutrients and space, especially when they are in pots. Therefore, they should be fertilised once a year. Perennial plants should be fertilised at the end of Winter with a mix of compost and manure so that the plant can get all 17 nutrients it needs.

Manure should never be in direct contact with the root because the excess of nitrogen in it would "*burn*" the root and prevent a balanced entrance of the other nutrients, thus, the plant would suffer from a temporary deficiency of nutrients. If inorganic fertilisers are added, it is vital to confirm that they contain the three macro nutrients that the plant requires in largest amounts: nitrogen, phosphorus and potassium. In

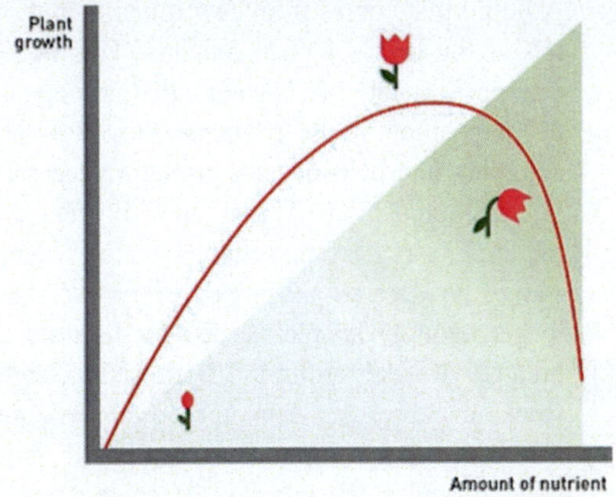

Fig. 3.7 Law of minimum. *Illustration* Juan Pablo Bruno

Lima, parks and gardens are usually fertilised with chemical fertilisers; this type of fertiliser kills the fauna in the soil which is vital for the decomposition of organic matter that ensures the presence of nutrients in the long term, so the plant becomes dependent on chemical fertilisers.

The most frequently used combination of synthetic fertiliser is urea, superphosphate and potassium nitrate or a compound fertiliser with these three elements. If so, it is convenient to use three doses of nitrogen in the form of urea, since it is an element that dissolves easily and is absorbed by the plant without any problem. However, when too much nitrogen is used, part of it will go to the soil and end up in the phreatic zone, deteriorating soil fertility and aquatic ecosystems. On the other hand, phosphorus needs only two doses because it is lost slowly as well as potassium which only needs one doses at the beginning of spring.

The second doses of nitrogen and phosphorus should be applied before blooming and the last doses of only nitrogen is applied after the harvest. Only when inorganic fertiliser is used the first three times, it is convenient to fertilise a fourth time, but with micronutrients to ensure the proper development of the plant. This last fertilisation should be foliar, and it should be done at the beginning of the blooming stage with magnesium and other microelements. When the yearly fertilisation is done with organic fertiliser, this foliar process is not needed since organic fertilisers contain all the nutrients required by the plants. Plants with a short vegetative period, such as flowers should only be fertilised once in their short vegetative period, while they are small and only with nitrogen since they grow very quickly.

Before placing the plants in a garden, it is important to prepare the ground, especially considering that the plants will spend decades in the same location.

In order to prepare the ground for trees or bushes, a 50 × 50 × 50 cm hole should be dug and organic matter such as decomposed manure should be placed at the bottom covered by a layer of plant soil or worm castings and soil in order to prevent the roots from being in direct contact with the fertiliser. Once this process is concluded, the plant can be placed. When doing so, the root crown should be at ground level and soil added should be packed.

To prepare the ground for planting we need:

- farming soil
- river sand
- compost
- Earthworm castings
- moss
- horse and cow manure.

Soil is not only a source of nutrients for plants, but it also gives the plant support, airing it and retaining humidity. Each one of its components gives the soil certain properties, that combined in different proportions make it better for some plants than for others according to each specie's need for nutrients as well as degree of soil acidity and salinity. Since roots need to develop underground, the soil should be loose enough especially in the case of annual flowers.

In this way:

- **Farming soil** enables the root to anchor properly, besides having the required nutrients for plant growth. Based on origin, the amount of sand or clay in soil can vary. The more sand it has, the better it will drain. However, if there is too much sand, water retention is difficult so the plant will be dependent on irrigation and will suffer from draught very quickly. If the farming soil has too much clay, it will need sand so that the water does not become stagnant in the ground and cause the plant to rot.
- **River sand** loosens the soil enabling water to flow easily and keeping roots aired. However, it is a poor source of nutrients so it cannot be used without other components. It is usually used in plant nurseries since they do not depend on nutrients and for flowering plants which have a short vegetative period.
- **Moss** comes only from plant matter and it is not totally decomposed. Therefore, it is rich in nutrients that cannot be assimilated directly by the plant, since it requires a previous decomposition process. In this sense, nutrients remain for a long period of time and they are not lost with irrigation water since moss retains it easily. In fact, it is only recommendable for plants that tolerate constant humidity of the soil. Furthermore, moss acidifies the soil slightly. It is ideal for indoor plants and for plants that come from the tropical rainforest.
- **Cow and horse manure** are so rich in nutrients that they might cause a temporary nutrient disbalance in the plants called "*burning*" if applied directly. Manure also acidifies the soil producing a disbalance in the nutrients available. This temporary disbalance occurs because the nitrogen in the soil is used by the microscopic fauna that decomposes the manure and only when it dies can the components be used by the plants. Therefore, before its application, it is convenient to prepare the manure leaving it in the open air for a couple of weeks, watering it continuously until it stops releasing its distinctive unpleasant smell in the same way as compost.
- **Compost** is rich in macronutrients as well as micronutrients and it is easily assimilated by the plant. Compost is formed by decomposed organic matter. It can be prepared at home or in the city park with the remains of dead plants.

To prepare compost, organic remains from waste are used alternating with one layer of manure. Additionally, compost can be made with the remains of a harvest alternating it with manure, placed on the ground or in a hole. A liquid that comes from cow rumen is put on top of the layer of plant remains. This liquid contains a lot of bacteria that accelerate decomposition; it also contains ashes. Both are acidifiers and help develop bacteria necessary for decomposition. Water and oxygen are required for decomposition, so the layers should be watered and mixed every two days to get enough oxygen needed for bacteria to breathe. It takes compost from six to eight weeks to be ready, depending on temperature. To ensure that compost is decomposed, temperature and smell is tested. If compost is at room temperature and does not smell like manure, it is decomposed and can be used without fear. If not, it is convenient to wait a couple weeks, watering it and moving it until it is ready.

- **Earthworm** castings come from compost which is placed on the ground alternated with soil. There should be at least 5 layers before placing the earthworms on top of them so earthworms can eat the compost and as they digest it, produce waste very rich in decomposed nutrients that are already available for the plant. Earthworm castings can be used to fertilize a garden every three months.
  Earthworm castings are more deeply decomposed than compost since they have been through the digestive system of the earthworms so they are a complete source of nutrients that can be used directly when planting. Furthermore, they are aerated and have good drainage since they form small balls that enable water to flow easily. If earthworm castings are not properly prepared, small pieces of plants will be in it. In this case, it should be combined with the other nutrients.

3.4.1.1 Soil Mix

The soil mix should be designed based on each plant's requirements and place of origin. However, it is important to consider that, according to our prior explanations the mix should be renovated since plants feed from the nutrients in the soil in order to grow. Therefore, it is also important to place together plants with similar agronomic requirements.

Soil mix adequate for the plants

- Indoor plants mostly come from the jungle and require soil with a high percentage of moss. In fact, many can grow perfectly only in moss.
- Since trees and bushes are perennial, they require a high content of matter for reserve. Therefore, soil should be prepared with compost and manure. Plants should be fertilised every year in August with an equal amount of compost and manure.
- Tropical plants require acid soils with a high content of nutrients as well as dark soils with high retention of humidity. Therefore, the soil mix could be composed by compost, moss and humus.
- Flowers require light and sandy soil. They can be fertilised with an equal amount of farming soil, sand and earthworm castings.

In Lima, these soil mixes should be used at least once a year in August at the end of Winter. Soil mix should be added to indoor plants at least three times a year. So that plants grow well it is vital to place together only plants with similar requirements of soil and water.

Soil colour can help us know what nutrients it has. When soil is dark, it usually has a high content of organic matter, whereas, lighter soil with a sandy or dusty colour has little organic matter. The best soil has an equal amount of sand, clay and silt and is given the name of loam. When this soil is moist, it forms a lumpy structure on the hand (small separated lumps), On the other hand, if the soil is too sandy, it does not form lumps and if it has too much clay, it forms a compact mass like the one use to make pottery. The last two do not allow the flow of water so the plant rots.

3.4.2 Water

Up to 80% of the wet weight of the plant can be water and in succulent plants it can be up to 95%. Water is absorbed by the plant roots and is evaporated through the stomates, which are specialized cells located in the leaves.

Water in plants has the following functions:

- Regulate plant temperature,
- Allow the intake of nutrients,
- Dilute nutrients,
- Regulate the opening and closing of stomates,
- Eliminate excess of salt.

Water in the plant plays an important role of refrigeration. By means of evapotranspiration water is released from the stomates in the form of vapour, then, water is absorbed again by the roots together with nutrients in the soil. In the plant, water moves away from zones with a high concentration of nutrients diluted in water to one with a lower concentration. It travels against gravity, from the bottom to the top. The absorption of water through the roots is due to an exchange of equivalent valences called ionic exchange. Once the water is absorbed, it travels through intercellular spaces, following the flow of evapotranspiration, left by the elimination of water through the stomates; it goes from the lowest to the highest pressure becoming available to the zone where the nutrients are needed. There, water is absorbed by the cells when hydrogen (H+) and hydroxyl (OH−) ions are exchanged with other ions in the intercellular spaces. If the plant does not get enough water, it will not be able to feed properly given that with no water there are no nutrients. In severe draught a plant famishes.

Plants resistant to draught which causes an excess of salt in the soil have different defence mechanisms. Some plants might not have ionic exchange in the root, letting all nutrients in without selection to later deposit them in the vacuoles. Therefore, their number of vacuoles in the leaves increases in order to store salt. They also shed their leaves often to eliminate the absorbed salts. We should also remember that potassium helps fight draught since it regulates the opening and closing of stomates. Under these conditions, plants will not open their stomates as frequently as under regular conditions. Another mechanism to fight draught is the modification of photosynthesis found in cacti, typical of desertic ecosystems. These plants prevent excessive water evaporation by only opening their stomates at night. It would be impossible to survive under these conditions if the plants did not gather water in their trunk to prevent it from overheating at night.

Plants need clean leaves to prevent stomates from clogging. In rainy, hot and saline zones, plants do not only eliminate water through the stomates but also excess salt that could be toxic otherwise. Sometimes, leaves are cleaned with milk, oil or even beer to make them bright and shiny. However, this can clog the stomates, decreasing the exchange of vapour and water.

3.4.2.1 Irrigation

The frequency of plant irrigation depends on the soil it is growing in and, overall, on the drainage it has. When soil is sandy the plant requires more water than when soil is clayey or compact. On the other hand, a mix with moss or organic matter increases water retention and decreases the need for water. It is important to remember that plants also require air to survive and they also breathe from their roots, so too much water could cause their roots to rot, specially in clayey soil.

The size of the root is determined by the plant's genetics, type of soil and frequency and form of irrigation. Roots develop depending on water availability, moving towards the water source. Therefore, when a plant is watered frequently with enough water, the roots are smaller than if they are watered sparsely. When water is scarce, roots look for water in the underground or move towards water or sewage pipes.

Less than 80% of irrigation water is used by the plant. Most of it is lost by gravity; another part of it clings to particles so strongly that the plant does not have the strength to absorb it so finally, only a small portion goes to benefit the plant. Once a plant has grown and adapted to a system of irrigation, this system should not change, as the plant will suffer to adapt to a new one; this process implies the relocation and development of roots. If the plant is fully grown, the process is very difficult, and the plant might perish.

To prevent the root crown from rotting, water should not be in direct contact with it. Neither should moss or compost as water retention will cause the crown to rot. If water has been previously stored and has been left stagnant, we should prevent salts and calcium from accumulating in the base, thus, affecting the plant. This is especially important for indoor plants that require a lot of care such as the African violet *(Saintpaulia ionantha)*, the gloxinia (*Gloxinia perennis*) or bulb plants; these plants should be watered by imbibition, submerging the pot in water so that it flows from the bottom to the top.

3.4.3 Light

Plants require light for photosynthesis as well as for performing vital functions. With daylight, a plant transforms solar energy into sugar. To have enough energy for this transformation, it absorbs carbon dioxide, water from the soil and light. As a result of photosynthesis, the plant has the energy it requires to live and release oxygen through the stomates and purify the air. The plant breathes 24 h a day and absorbs oxygen to produce energy it needs to elaborate nutrients; at the same time, it releases carbon dioxide into the air. At night it only breathes and releases carbon dioxide. On the other hand, during the day it releases more oxygen than carbon dioxide. In average, in 24 h the plant releases more oxygen than carbon dioxide since oxygen is produced with the liberation of energy during metabolic processes and the carbon dioxide released is used for photosynthesis.

The plant has special organs to transform light energy. These organs are called chloroplasts located in the leaf cells. The number of chloroplasts can vary depending on the plants and the climate. Herbaceous plants and exceptionally bushes or vines may have chloroplasts on the stems.

3.4.3.1 Chlorophyll Synthesis

Chlorophyll is synthesised at a visible wavelength, the spectrum of the light absorption depends from the type of chlorophyll. When the plant does not get enough light, internodes elongate, and leaves become yellow or white as well as smaller. Besides chloroplast, plants have other pigments called chromoplasts which can also give colour to the leaves, and less efficiently, can help capture light at different wavelengths. Leaves of specific colours have modified chloroplasts and can even change form. Under these conditions, chromoplasts step in producing a red or yellow hue or combines both hues to make orange. If the plant has chloroplasts and chromoplasts, it may be purple.

Plastids called leucoplasts are colourless and give a white appearance to the plant's organs enabling the storage of starch; this is the case of potatoes and bulb ornamental plants.

The number of chloroplasts in a plant varies depending on the species and the amount of light that the plant gets. The plants can adapt to the changes in intensity of the light, varying the number of chloroplasts. The greener the colour of the leaf, the larger the number of chloroplasts and, therefore, the more capacity the plant will have for photosynthesis and will be able to grow better in the shade. There are several types of chlorophyll which differ in form and amount of light absorbed expressed in a variety of colours. The difference of plastids depends on their internal form which allows light to be captured. In this sense, there are two types of chlorophyll.

- Chlorophyll A: present in all plants.
- Chlorophyll B: present in most green plants.
- Carotenoid Pigments: carotenes and xantophylls, present in chromoplasts. They are in the yellow and red spectrum depending on whether they have oxygen or not respectively, Therefore, their colour could change when the amount of oxygen varies. They are in most plants.

A change in the intensity of light can also affect the colour of the leaves so plants can transform and lose their distinctive qualities, but transformation can take some time. Therefore, a plant should not be taken abruptly from a shady spot to a spot with a lot of light since it could cause leaves to burn when there is too much light. On the other hand, if light decreases abruptly, the plant will suffer from a deficiency of nutrients and could even perish if the situation is prolonged. When we change a plant from a sunny place to a shady one, it requires a period of adaptation to increase its capacity to do photosynthesis. During this period the plant will suffer from a deficiency of nutrients. However, because plants do not have the same number of chloroplasts, they will adapt to change differently.

Plant adaptation in response to changes in light

- Light coloured plants or ones with white or yellow spots have difficulty adapting to low intensity light whereas plants with dark green or reddish leaves adapt more easily. Plants with white hues or spots become totally green when there is a deficiency of light, losing their distinctive characteristic.
- Grey, silver or brown plants can be planted under trees since most of them come from the tropical rainforest and are accustomed to little light and a lot of moist. Silver leaves are so because they are coated with a wax to prevent rotting. These plants have a larger number of stomates, so they have better absorption and circulation of nutrients if atmospheric humidity is high.
- Red and violet plants need a lot of light. In the shade, they become green and the number of chloroplasts increases, since there is not enough light to cover for the needs of chromophile in charged, in this case, of photosynthesis.
- Plants with specific morphological qualities like cacti which has wax on its trunk and leaves transformed into spines, or succulent plants with retaining leaves have adapted to live under the burning sun. The silvery colour in the wax prevents the sun from burning excessively. Under low intensity light the plant elongates and might even grow leaves instead of spikes, like some types of cacti.

3.4.4 Temperature

Plants can grow faster in higher temperature, yet, it cannot be too high. However, there are plants that cannot adapt to heat due to their genetics determined by natural selection in their place of origin. The temperature required by the plant is determined by the plant's number of stomates which defines evapotranspiration, speed of photosynthesis and water intake, together with nutrient selection that each plant performs according to their species. Transpiration does not only depend on temperature but also on the plant's age and fertilisation.

Therefore, for practical purposes we can classify plants based on place of origin in tropical, subtropical and cold zone plants. Temperature ranges from 45 to 30, 30 to 10 and 10 to 0 ° C, respectively (Smith and Smith 2007).

3.4.5 Air

Plants require air both as carbon dioxide (CO_2) for sugar production, during the day, and as oxygen (O_2) for respiration, all day and night. Thus, plants liberate energy, decomposing sugars produced during the day by photosynthesis and using these elements to make substances required by the plant.

Photosynthetic reactions occur in the chloroplasts in the cells producing sugar, water and oxygen. In order to do this synthesis, the plant requires solar energy, water and carbon dioxide. Respiration reactions occur all 24 h, in the cell, more precisely in the mitochondrion, finally producing carbon dioxide and water.

Plant respiration depends both on the concentration of oxygen, and on respiration velocity determined by temperature. The lower the temperature the slower the reaction. Therefore, to preserve flowers, fruit, seeds or parts of plants, they are placed in cold atmospheres with a low content of oxygen, delaying respiration and, consequently, the aging process.

The radicular system also breaths. When soil is not drained properly and the amount of oxygen is low, breathing becomes difficult. This could cause anaerobic reactions and the production of alcohol which could make the root ferment and rot. First a cell dies, then the root and finally the entire plant.

If there is a deficiency in the amount of oxygen, the plant will grow rickety due to a deficiency of nutrients absorbed by the root. That is why, plants appropriate for muddy soil which are referred to as neumatophors, produce fake roots called adventitious roots. These are stems that grow towards the soil and anchor in it, enabling plant stability when roots cannot develop as well as air absorption. For this to occur, there should be a change in the flow of hormones to allow the stems that move towards the soil to have negative phototropism (see chapter on physiology).

3.5 Physiology

3.5.1 Photosynthesis

The objective of photosynthesis is to accumulate energy in the form of sugars. Nutrients elaborated in the leaves are transported by the phloem to other plant organs. Based on ecological conditions, plants have developed three types of photosynthesis according to their photosynthetic efficiency; they can be classified in C3, C4 o CAM. C3 plants make sugars with 6 carbons, C4 plants produce sugars with 4 carbons and CAM plants produce sugars with 3 carbons; the latter are also referred to as plants with crassulacean acid metabolism, only opening their stomates during the night and, consequently, processing sugars more slowly.

Photosynthesis in **C4 plants** produces fructose, a sugar with 4 carbons, which gives C4 photosynthesis its name. In comparison to other types of photosynthesis, this photosynthetic process is more efficient because it forms sugar faster, enabling plants to get maximum benefit from light intensity in the ecosystem. C4 plants fixate carbon dioxide (CO_2) and water in a 4-carbon compound (oxaloacetic acid), forming fructose. These plants can be distinguished from others by their elongated leaves. In this group, we can find several types of grass abundant in prairies or in our high-Andean grassland ecosystem as well as tropical plants in dry zones. An example of C4 plants are the ravenala (*Ravenala madagascariensis*), the bird of paradise (*Strelitzia reginae*) and all types of grass from tropical and subtropical zones.

C3 Plants Fixate Carbon Dioxide and Transform It into a 3-carbon Compound (Phosphoglycerate) Which is a Temporary Product of This Form of Photosynthesis and Gives C3 Plants Their Name. The Final Product of Photosynthesis is Glucose, a Six-Carbon Sugar. To Make Glucose the Plant Needs More Energy and Time Than to Make Fructose. Therefore, These Plants Are Less Efficient in Ecosystems with High Intensity Light, not Benefitting from It at the Same Speed as C4 Plants. C3 Plants Have a Medium Photosynthetic Capacity and Moderate Growth Per Unit of Time.

However, the sugar formed (glucose) has a larger number of bonds so energy is stored more efficiently. Most plants are in this group and they grow in temperate ecosystems. Such is the case of the rose or the hydrangea (*Hydrangea macrophylla*). However, some grow in tropical humid climates like the umbrella tree (*Schefflera actinophylla*) and the flamingo flower (*Anthurium andreanum*). These plants have round or oval leaves.

Unlike the above mentioned, **CAM plants** open their stomates to allow absorption of gases only at night. This way they save energy and water because while stomates are not open, there is no exchange of gases nor loss of vapour. At night, these plants fixate carbon dioxide and store it in their cell vacuoles, leaf organelles that enable the storage of chemical substances. Acid is stored in form of malic acid or malate ($C_4H_6O_5$) which is easily metabolized. During the day, malic acid is transformed again into carbon dioxide and water. Then, it is transported from the vacuoles to the chloroplasts where light energy is fixated, and sugar is produced. This photosynthetic path is slow which is the reason these plants grow slowly. The advantage of this process is that it prevents water evaporation since the plant stops the release of water through the stomates during the day. That is why these plants are perfect for desertic ecosystems where there is little water. Cacti and succulent plants use this kind of photosynthesis, saving water by preventing evapotranspiration. Leaves either have a lot of water stored or have transformed into spines leaving photosynthesis to the trunk.

Due to these physiological differences, C4 plants are the fastest to grow; most tropical plants are in this group. CAM plants are specially adapted to grow in the desert since they open their stomates only at night; cacti are in this group. C3 plants are mostly found in temperate zones and their growth is moderate but sugar storage is very efficient.

3.5.2 Plant Hormones

A plant hormone is a chemical substance that in small doses modifies the plant's growth and development processes promoting or inhibiting them. Hormones are produced in specific parts and are transported to the rest of the plant. Even though hormonal production is determined genetically, it can be activated or inhibited by the environment. The main environmental factors that trigger hormonal changes are temperature and light. The plant modifies its hormonal composition throughout its lifecycle based on its stage of development as well as environmental conditions. Plant hormones are: gibberelline, auxin, abscisic acid, cytokinin, and ethylene. Hormonal reactions are complex and are subject to a balance of these hormones since the doses of each in a plant can inhibit or promote processes.

Auxin, in balanced doses, produces cell elongation reflected on sprout and plant growth. However, in excess it prevents cell elongation. Auxin is synthesised in the apexes and from there it is distributed to the rest of the plant by the parenchyma that surrounds the vascular bundles. Its effect depends on the doses; low concentration of auxin has an effect on the root, whereas more concentration has an effect on the terminal bud, yet, too much prevents cellular elongation. In general, plants have more auxin during the first stage of growth.

Auxin produces apical elongation, inhibits growth of lateral branches and induces sprouts to bend towards the light due to cellular elongation in order to reach the light when its dark. Apical dominance is concentrated in the terminal sprout called coleoptile, both in the root and in the terminal sprout of the plant. For this reason, when the coleoptile is removed, the apical dominance is interrupted. This fact is important in the city where the final size of the tree should be reduced. Some plants have to compete for the light in their early years, such as the ones that grow in the rainforest under an infinite number of adult trees, so tall that they do not let the light go through. Under these conditions, auxin assures the survival of these trees allowing them to win the competition for light by helping them grow fast. This is the case of palm trees which grow very fast vertically in their first stage of growth.

Horticulture also considers this function for plant propagation. An incision is done in the bark, damaging the cambium, thus interrupting the flow of auxin and concentrating it in the tree incision promoting, then, radicular growth of the plant. For this reason, the propagation of woody trees is possible with cuttings or air layering.

Another effect of auxin is the formation of a callus which is an unorganized mass of cells in the cambium (parenchyma) that divides vascular bundles. The callus is vital for the plant to heal a wound. Auxin, which promotes cellular elongation together with cytokinin which promotes cellular division enable regeneration of damaged tissue. This is not possible in herbaceous plants because they have no cambium. Even though auxin stimulates the development of roots and, therefore, promotes the growth of adventitious roots. Effects on the roots depends on the doses. In low concentration, it stimulates the production of lateral roots and in high concentration it blocks this production.

Auxin also acts on seeds and fruit by stimulating their development and preventing deformation or abortion. The seed promotes auxin production after fruit fertilisation. If the fruit has not been fertilised, it may start developing but it will be aborted prematurely. This occurs with *cocktail* avocado, which never develops fully since it has no fertilised seed. A similar effect takes place when only some seeds of a multiple fruit have been fertilised, therefore fruit develops unevenly which means one part of the fruit will remain developed while another will remain smaller, such as the fruit of chirimoya (*Annona cherimola*).

Auxin prevents the flowers, fruit and leaves from falling, thus, plants with high amounts of auxin may keep their fruit for a long time. Such is the case of the avocado tree (*Persea american*) whose fruit stays on the tree for several months if they are not harvested. Auxin can also benefit flowers of monoecious plants and help produce larger harvests of, for example, squash (*Cucurbita maxima*) by giving the male flowers which do not produce fruit, a higher level of auxin externally, thus, making them female. In this way, the yield of squash will increase.

Artificial auxin can be produced in a laboratory under the name of indoleacetic acid (AIA). Auxin powder is used for the reproduction of plants from cuttings. First, the cutting is submerged in water and later it is placed in auxin powder, ensuring the powder adheres to the wound to form the callous and roots. In this way, you get a new plant through asexual reproduction which is a clone of the plant the cutting was gotten from since it has the exact genetic characteristics.

Gibberelline causes cell elongation and division. There is a larger amount of this hormone in the juvenile stage of the plant. It promotes sprout and internode elongation and development as well as seed germination in plants that require light to germinate. Gibberelline promotes flowering in response to long days and is responsible for the formation of male flowers in monoecious plants. Together with auxin, gibberlline promotes parthenocarpy or production of fruit without fertilisation but by cellular division. The production of parthenocarpic fruit in horticulture yields seedless fruit, such as banana, fig (*Ficus carica*), pineapple. This type of fruit has more pulp and are very nice to eat. However, this hormone has a negative effect on the development of lateral roots.

Cytokinin is a plant hormone that promotes the development of side branches. When there is more cytokinin, lateral buds are released from apical dominance caused by auxin. The cytokinin/ auxin relationship determines cellular division and growth. This hormone is typical in the reproductive stage of the plant and delaying its aging process. Cytokinin also promotes the development and cellular division of the chloroplasts. Cytokinin interacts with auxin and intervenes in the replication of cellular DNA.

Abscisic acid (ABA) acts as a growth inhibitor, modifying the effect of other hormones. It increases in short days causing leaves, flowers and fruit to fall, inhibiting sprout growth and causing bud dormancy. It is vital for plants that come from places with cold climate. It allows buds to rest during dormancy preventing the plants from having sprouts before time. Abscisic acid controls respiration of the stomates, regulating the amount of water in the plant as well as the osmotic pressure which allows the plant's survival in dry climate.

Ethylene is a gas hormone that stimulates fruit to ripen and acts in the last phase of abscission causing flowers and fruit to fall after the phase in which auxin works, when the fruit is already ripe. It also acts in young fruit causing it to fall if there is no auxin due to lack of fruit fertilisation. It stimulates germination and growth of grains, bulbs and cuttings. It provokes radial growth and reduces elongation. It stops radicular growth promoting the formation of lateral roots, radicular hair and adventitious roots. Furthermore, it stimulates the formation of airborne water plants by accumulating gas in the leaves. Finally, it affects the formation of female flowers in cucurbitaceae and pineapple. It is used a lot in ripening tropical fruit. Since it is a gas, it is easy to use, and it is the reason that wrapping fruit like avocado (*Persea americana*) in newspaper makes it ripen because ethylene is concentrated around the fruit preventing the gas from dispersing.

3.5.3 Photoperiod

Most plants require a certain length of continuous darkness to bloom; this capacity to regulate blooming is called photoperiod which means that the plant responds to a certain length of continuous darkness throughout 24 h to bloom. Photoperiod is determined genetically, and it is part of the plant's evolution process which enables a species to live under certain conditions of light. In this process, the protein called phytocrome plays an important role. The red light (day light) and far red (night light) is detected by the phytocrome and the amount of phytocrome produced during daylight determines when the plant will bloom.

Based on photoperiod, plants are classified in:

- **Long-day plants**: These plants need to accumulate a certain number of days with more than ten to fourteen hours of light to be able to bloom, and they do so in summer. The number of hours accumulated of darkness determines flower formation. An example of long-day plants are the carnation (*Dianthus caryophyllus*) and the snapdragon (*Antirrhinum majus*). In these plants, blooming is promoted by blue light photoreceptors and activated by the plant hormone gibberelline.
- **Short-day plants**: These plants need to accumulate a certain number of days with less than ten to fourteen hours of light to be able to bloom, and they do so in winter. An example of this is the rose. Phytocrome promotes blooming in short-day plants when nights are long. If they are not, plants will not bloom. Short-day plants are the chrysanthemums (*Chrysanthemum morifolium*), the African violet (*Saintpaulia ionantha*), and the Gloxinia (*Gloxinia perennis*).
- **Neutral-day plants**: These plants bloom any time a year and do not respond to hours of light. This means, they have no sensibility to photoperiod. Neutral-day plants are the bird of paradise (*Strelitzia reginae*) and the geranium (*Pelargonium hortorum*).

Long-day plants can make up for their need of light by interrupting the night. This is done in floriculture to make long-day plants flower. On the other hand, short-day plants need the night to continue which could be done just by putting the plant in the dark. This means, blooming is determined not by the length of the day, but by the continuity of night. Photoperiod also determines phenomena like bulb and tuber formation, both in combination with changes in temperature.

The native plants of Lima's ecosystems are day-neutral, because in the subtropical zone in which Lima is located the seasonal difference in daylight hours is minimal.

3.5.4 Thermoperiodism

Thermoperiodism is the response of plants to temperature. It is essential for the plant's survival in ecosystems with cold winters. However, it does not occur in plants that come from warm ecosystems. It can be described as the number of hours a plant should be below a certain temperature to modify its growth stages. It enables plants to bloom and germinate in cold zones.

Thermoperiodism is not an immediate response to changes in temperature. Each plant needs a certain length of time below a specific temperature to modify its growth stages. This length of time depends on the plant's genetics as well as the temperature at which the time will start running. Once time has elapsed at the mentioned temperature, the hormonal changes necessary for the plant's growth stage to modify will occur.

Thermoperiodic response is typical of plants that come from cold climates and ensures that plants do not bloom or sprout when it is cold which could cause the death of the sprouts as well as the plant.

Each organism should be at different temperatures for a length of time to produce changes in its growth phases. Once the required number of hours elapse, the respective plant hormones will activate to produce the necessary hormonal modifications in the plant. Most bulbs such as tulips (*Tulipa gesneriana*), require a period of cold to go from the vegetative stage to the productive stage. Other plants like apple trees (*Malus domestica*) or pear trees (*Pyrus communis*), also have specific requirements. For example, a pear tree from the Anna of Israel variety needs the temperature of the Celicious of Viscas and other varieties of apple. Thus, for example, apple varieties with low chilling requirements, such as the Anna and Delicious de Viscas, are among the few that can bear fruit in Lima's climatic conditions.

3.5.5 Dormancy

Plant dormancy refers to a stage of rest stimulated by a decrease in temperature and light, which regulate the activity of abscisic acid (ABA), a plant hormone that increases when days are shorter. These two environmental conditions trigger a decrease in plant activity that leads to hydric stress since ABA causes a reduction in the opening of stomates and, therefore, a decline in plant's nutrients.

ABA acts by inhibiting the effect of other growth hormones. This effect occurs in both cold and dry ecosystems. Abscisic acid also causes trees to shed their leaves in autumn, just before trees go into their first period of dormancy. The shedding of the trees is vital so that the nutrients in the leaves can relocate in the trunk and prevent it from falling.

Dormancy is typical in plants from cold ecosystems. It helps them survive winter and frost. During frost water becomes solid and increases volume. Cells are especially sensitive to frost since their size increases and they break the cellular wall causing the cell to collapse. Leaves are more exposed to cold than the trunk that is protected by lignin.

Lima's native plants do not have marked thermoperiods or photoperiods since the temperature varies little throughout the year. In turn, in ecosystems with dry and rainy seasons—such as Loma, located on the city's periphery—native plants like the amancay (*Ismene amancaes*) accumulate water in their bulbs to survive in the dry season, during which the time they enter a state similar to dormancy.

3.5.6 Tropism

Tropism refers to the plant's response to stimuli, yet, geotropism refers to the plant's response to gravity. Just like any object, plants are affected by gravity which has an effect on hormonal distribution and, thus, on the growth and development processes. Plants detect gravity with the amyloplasts which are plastids with a content of starch. These are concentrated in the root cap and in the stem apexes. Auxin is the hormone responsible for the response to tropism and it concentrates in the root cap and the stem apexes. It is also present in the germination process of most seeds. That is why we should cover the root with soil so that it can germinate.

There are different types of geotropism based on the plant's response to light. Then, geotropism can be negative or positive. Whereas the roots have positive tropism or geotropism, growing towards gravity, leaves have negative geotropism or phototropism growing towards the sky. Geotropism is detected by auxin which is in the coleoptile protected from the light since the hormone is synthesized in the dark. Geotropic reactions depend on the plants stage of life, environment and specie genetics.

If the plant encounters a chemical in the soil which will impede growth, it will react staying away, whether these are chemical substances or changes in the soil's pH. This is called chemotropism. In this case, the plant reacts to the ions in the soil and keeps away from them or moves towards them. This reaction is essential for survival in order to avoid acids or an excess of salt in the soil as well as to find nutrients.

The plant also reacts to substances like water or air. The response of a plant to water is called hydrotropism whereas its response to air is aerotropism. Finally, plants also react to solid substances staying away from them; this is the case of rocks, other roots, constructions or simply compact soil. This response helps the plant find the nutrients it needs for its development. Root growth is reflected on crown growth since the plant should keep the crown and the root balanced. In the city, we can observe that plants grow in accordance with pipes or solid elements in the ground.

3.6 Most Frequent Pests and Disease in Ornamental Plants

In the city most ornamental plants are strong and resistant dominant plants with big and flamboyant flowers that in their place of origin would displace others. Here, plants are challenged by soil compacting caused by pedestrians, ground water contamination produced by sewages and many other struggles. The advantage is that cities are slightly warmer than the countryside. However, in Lima this is a disadvantage.

Another limiting factor in the city affecting adequate growth of plants is air pollution. When a plant is in a place with high levels of air pollution, it is more prone to pests and disease, since their stomates are covered by pollutants. This prevents them from functioning normally, thus, reducing photosynthesis activity and affecting plant growth and defence. However, plants can be classified based on their tolerance to pollution.

Plants that have had to struggle to survive in environments with a lot of competition or with too many difficulties are definitely more resistant to pests and disease as well as to lack of adequate conditions of air, water and soil. This will depend a lot on the process of genetic selection the species went through in their place of origin. The more similar the environment is to its place of origin, the better the conditions will be to grow and resist pests and disease.

3.6.1 Pests

A plant pest is any animal that damages the plant. Most are insects that feed on plants. Snails, slugs, mites, nematodes and sometimes birds are pests. Insects are very resistant to adverse environments. Their high reproduction rate, short lifecycle, small size, provide them with a great capability to adapt.

Based on their mouths, insects can be biters, lickers, stinger-suckers and suckers. Their speed to devour the plants will depend on this. Most insects go through metamorphosis changing from larva to pupa and from pupa to its adult form. This makes them harder to spot and control. The larvae have a great capacity to feed and can quickly destroy a plant.

The presence of pests depends on climate, resistance and competition in the environment where plants are growing. Warm climates promote insect proliferation. The introduction of a plant frequently makes it easier for pest to spread more quickly since it has no natural enemies. When the plants face pest, the plants act in group producing volatile substances that scare insects away after plants are attacked. The resistance to pests of plants subject to insecticides is lower because the insecticide prevents the identification of the volatile substances released by the plants for defence. These substances, synthesized by hormones can be transferred by the mycorrhiza (Babikova et al., 2013, cit. Wohlgemuth et al. 2019: 104). However, there are no mycorrhiza in plants fertilised with nitrogen. Plant resistance in the city is low since plants are subject to chemical substances and pollutants; the city could even be subject to draught like in Lima. This low resistance produces a vicious circle since it increases the need for fertilisation with chemical products.

The most common everlasting pests are called key pests. As follows, there is a description of the main key pests in Lima.

- **Whitefly (Aleurothrixus floccosus)**:
  The whitefly is a sap-feeding hemipteran insect. It is found on the leaf underside where adults lay their eggs. Larvae feed off leaf sap making a part of leaf fall and causing sooty mould which always accompanies whitefly.
  To control whitefly, we should consider its lifecycle. Eggs laid on the leaf underside hatch in two weeks releasing voracious larvae. Under the environmental condition in Lima, adults live for about two weeks. To ensure the elimination of larvae before they reach adulthood and lay more eggs, insecticide should be applied twice with a two-week gap. The whitefly can be controlled naturally with tobacco tea. It can also be controlled with a solution of dish soap and water, made up with a bar of odourless soap (¼ kg) dissolved in 8 L of water.
- **Pea leaf miner (Liriomyza huidobrensis)**:
  It is a two-winged insect from South America. It is most harmful in its larva stage. The larva makes tunnels between the lower and

upper epidermis to feed on the leaf. The larva is 2.5 mm. long. When the larva is fully developed it goes into the pupa stage in the leaf underside. The adult is small and black with yellow spots. They can pose a serious risk for young flowers and herbaceous plants like spinach. This pest is controlled with yellow plastic traps or painted cardboard covered in oil which attract the adult insects that stick to the yellow traps.

- **Cutworms**:
  These are lepidopter larvae in the Noctuidae family (night-flying moths). They feed at night, cutting down recently germinated plants at the collar. At daytime the 5-cm larvae are rolled up at the foot of the tree below ground and cannot be seen, they are rolled up. However, in the early morning all the young plants appear felled at the collar. These worms can be controlled by applying tobacco combined with eggshell at the foot of the tree. Another homemade recipe to control these chewing insects is placing traps submerged in a solution of 1 kg of sugar, ½ kg of molasses, half a bottle of beer and a shot of rum. Pieces of wood and cardboard should be submerged in the boiled solution and placed as traps in different parts of the garden. The traps should be collected and discarded once worms get stuck to them.
- **Snails and slugs**:
  Snails and slugs consume plant leaves eating the tender part and leaving only the midrib. They feed at night hiding in the day when it is warmer under the leaves closest to the ground. They can be controlled manually or by placing plates with black beer mixed with flour as bait so that the snails can fall in and drown. It is also useful to place ash around the plants. This will help keep snails, slugs and chewing worms away since ash adheres to their body and they cannot get rid of it.
- **Aphids (*Aphis sp.*)**:
  These are insects which mainly belong to the aphis genus. They are very small (about 1–2 mm), soft-bodied insects. They are usually light green. These insects suck sap out of young plant sprouts. Like the whitefly, when the aphids suck the sap out of a plant, they leave traces of saliva where sooty mould grows. Aphids are also virus vectors for viruses which weaken a plant. Plants attacked by aphids, sooty mould and virus are scrawny and their leaves are a yellowish green or they have yellow spots depending on the type of virus they have. Aphids can be controlled with tobacco tea in proportion of 400 gr of tobacco per 4 L of water. Another useful solution is made by adding 1 tablespoon of vegetable oil and 1 tablespoon of dish soap per each litre of water. The solution should be shaken before application. The oil will cover the insect and prevent it from breathing until it dies.
- **Red spider mite (*Tetranychus sp.*)**:
  It is a small, red or cream mite. It looks like a small 2 mm spider. It can be found on the leaf underside where it feeds on plant sap causing the damaged leaves to become a tan colour. Eventually, the tree will shed the leaves. It prefers tender plants since they are easier to chew. It is easier these mites to develop in dry and warm climate. They are controlled alternating plants with garlic or onion plants which act like repellents against the red spider. It can also be exterminated with solution of tobacco when there is only a small number of them or applying garlic tea in a proportion of 10 cloves per 2 L of water. Both solutions are applied once the water is cold.
- **Scale insect**:
  These insects are suckers. They adhere to stems and leaves. Females do not have wings and most of the time have a hard-shell-like covering which does not let it move. Winged males and nymphs move easily. Scale insects come in different colours. They can be brown, black or white. They are controlled with a solution containing one spoonful of oil per each litre of water with odourless soap (¼ kg.) in 8 L of water. The oil is shaken in the water before the solution is applied.

- **Ants**:
  Ants cut plant leaves and roots. They can leave plants bare. They are controlled when their nests are destroyed. This can be done by pouring hot water over them since it will force ants to move. Peppermint keeps ants away from crops. Peppermint tea can be poured on their nests or along the ant path.

3.6.2 Diseases

Disease in plants can be caused by different organisms: fungus, virus or bacteria. However, most disease in plants is caused by fungus.

3.6.2.1 Fungus

There are several types of fungus that can attack plants. Damage from fungus causes concentric circular leaf spots and rotting. Unlike fungus, bacteria cause humidity damage and often give off unpleasant smell. Both need humidity to develop.

- **Oidium (Oidium sp.)**:
  It is the most common fungus that affects ornamental plants. It can be found on leaves, flowers and fruit in the form of white powder. It is caused by too much humidity, It can be avoided by watering less and trying not to wet the ground at the foot of herbaceous plants like flowers or indoor plants with herbaceous stems.
  Fungus appears in humid seasons. In Lima, it can be seen in winter months. It is controlled by reducing humidity and increasing light which helps to keep the environment less humid.
- **Botrytis o Grey Mould (Botrytis cinerea)**:
  This fungus typically covers flowers and fruit with a greyish coating. It is common in cut flowers, roses and herbaceous plants. This fungus can be controlled by watering less as well as by not wetting flowers and fruit when watering. It is also helpful to sprinkle ash on the plants.
- **Stem rust (Puccinia graminis)**:
  This fungus commonly affects lawn grass, ichu and other types of grass. Fungus feeds on plants, leaving an orange, brown or black crust on the stem. It weakens the plant and delays development. It is common in places where plants cannot get proper nourishment and are not watered. It is difficult to control but its incidence can decrease when grass is cut timely and frequently.
- **Smut (Ustilago sp.)**:
  This fungus produces black pustules in grass seeds that can be detected when sitting on the grass since clothes will be spotted with a black dust from the fungus seeds. It is controlled by cutting the grass timely to prevent fungus seeds from forming.

3.6.3 Pest and Diseases Control

Pests can be controlled either chemically or physically. In organic agriculture physical control is carried out when pests are evaded through early harvesting before crops are affected. Biological control is carried out using pest's natural enemies. As follows we will describe different forms of control (see Tables 3.14 and 3.15).

3.6.3.1 Biological Control

It is about recognising the pest's natural enemies in order to release them so that they can feed on the pest insects. Commonly, natural pests are also insects called beneficial insects. As follows we have examples of biological control.

- Ladybugs: the use of ladybugs or insects in the Coccinellidae family which feed on aphids,
- Parasitic wasps: the use of parasitic wasps in the *tachinidae* genera to feed on butterfly larvae,
- Spiders: the use of spiders that eat insects,
- Bacteria *Bacillus thurigiensis*: when pest insects eat these bacteria, they consume the pest's intestines, killing them. They are used to control lepidopter larvae. They are also used to control beetles and dipterans (Martin

Table 3.14 Main pests in the city of Lima, own authorship

Main Pests in the city of Lima	
Common and scientific name	Natural control
Whitefly (*Aleurothrixus floccosus*)	Tobacco tea Spoonful of dish soap diluted in 1 lt of water
Mine fly (*Liriomyza huidobrensis*)	Yellow traps
Cutworms	Tobacco mixed with eggshell Traps submerged in 1 kg of sugar, ½ kg of molasses, half a bottle of beer and a shot of rum daubed on cardboard traps
Snails and slugs	Manual recollection Beer traps made with plates of black beer mixed with flour Ashes around the plants
Aphids (Aphis sp.)	Tobacco tea in a proportion of 100 gr of tobacco per 4 L of water Mix 1 spoonful of vegetable oil with 1 spoonful of dish soap per litre of water
Red spider mite (*Tetranychus sp.*)	Tobacco tea. Also, garlic tea in a proportion of 10 cloves per two litres of water
Scale insect	1 spoonful of soap, 1 spoonful of oil, diluted in 1 L of water
Ants	Hot water, peppermint tea

Table 3.15 Main fungi found in the city of Lima, own authorship

Key main fungi found in the city of Lima	
Common and scientific name	Natural control
Oidium (*Oidium sp.*)	Reduce humidity and increase light
Botrytis or grey mould (*Botrytis cinerea*)	Reduce watering and avoid wetting flowers and fruit when watering Sprinkle plants with ashes
Stem rust (*Puccinia graminis*)	Timely and frequent cutting of grass
Smut (*Ustilago sp.*)	Timely cutting of grass preventing seeds from forming

and Sauerborn 2006: 176). Bacteria strain can be found in stores to be spread around parks and gardens.

Plants can also be used to control pests biologically. Many plants in nature, around 2400 (Martin and Sauerborn 2006: 175), are repellents used to control pests. For example.

- **Tobacco**:
  Nicotine in tobacco is neurotoxic (Martin and Sauerborn 2006: 175). It is prepared by boiling the tobacco leaves and spraying the cold tobacco leaves onto the plants or grinding the leaves and sprinkling the powder onto the leaves. Tobacco controls larvae on leaves. However, it is also useful for hot-blooded animals and cutworms affecting its development (Martin and Sauerborn 2006: 175). Therefore, it should not be used excessively or when it is not needed.
- **Quassia or bitter wood**:
  It is a plant native from South America. It is used as an extract which is obtained by boiling the bark or wood. It is very potent, and it is used to control leaf larvae as well as lice and fleas in natural medicine. (Martin and Sauerborn 2006).

- **Avocado seed**:
  Tea made from avocado seed (*Persea amaericana*) is very effective to control beetle larvae called grubs. To make the solution, two avocado seeds should be boiled in a pot of water and once it is cold, it should be sprayed onto the plant which can be watered since the water will fall into the soil where larvae hide.
- ***Azadirachia indica***:
  The common name of this plant in Peru is nim. It prevents metamorphosis in insects (Martin and Sauerborn, 2006: 176). It acts in the pupa stage when the insect is going through a transformative process causing it to die of starvation.

3.6.3.2 Cultural Control

Cultural control is used in the countryside as part of farming activities to prevent or diminish the effect of pests. The main ones are:

- **Planting Season**:
  Plants should be planted when they grow best; in this way they will be able to withstand pests. This advice is especially applicable for flowers which are very sensitive it is specially useful for o pests and disease.
- **Crop Alternation**:
  A very common practice used in biological agriculture is to change the species of herbaceous plant cultivated. It is important not only to change species but also plant family. In this way, insects that remain after the previous crop will not find the plant they are looking for and, since most plants are specific, it will break the cycle. This is especially true for garden flowers and vegetables.
- **Mixed crops**:
  It consists of planting different species in the same space. This prevents weed from growing and if planned correctly, it can help control pests. Since there is a variety of species, pests do not find enough of the same type of food and do not prosper. This happens with plant beds as well as with bio-orchards.
- **Watering control**:
  Like plants, pests need water, some a lot, others only a little. Therefore, when the watering regime is changed, pests tend to migrate to other spaces. The same result can be attained by changing the type of plants used from one planting season to the next in case of flowers and herbaceous plants. So, when choosing the plants for the new planting season, it is important to consider that the they should have different requirements of water. As described lines above, the red spider mite likes little water whereas fungus develops when there is a lot of water. This method is also recommended when the area has been invaded by invasive plants that are difficult to eradicate like the nut sedge (*Cyperus rotundus*) or bamboo.

3.6.3.3 Mechanical Control

Mechanic control consists in using manual or mechanical methods with agricultural tools or machinery to fight pests.

- **Mechanical elimination of undergrowth**:
  Eliminating undergrowth helps control pests since it is often host to different types of pest. Undergrowth can host biter sucker insects such as larvae of aphids and butterflies (lepidoptera) and fungi.
- **Recollection of pest insects**:
  This type of control requires care and time; it can be done manually or with a pest vacuum. The latter can be done when the pest is big like the snail or some beetles. The vacuum can be used with big and small insects which are not strongly adhered to the plant such as the beetles.
- **Recollection of damaged fruit and dried leaves**:
  Dead flower, fruit and old leaves can host pests so they should be recollected and eliminated every week. This can be done

when plants are watered as well as during cleaning pruning.

- **Soil removal**:
  When soil is removed, beetle larvae that pupate under soil are destroyed. This can be done around the plant together with the cleaning pruning. However, it should not be frequent nor deep to prevent damage to the root and prevent fungus from entering. This process is frequently done to grass. When the soil is removed, it does not only control pests, but it also airs the soil triggering root growth as well as promoting the development of fauna in the soil.
- **Elimination of crop residue**:
  Crop residue has pests that stay in the fields awaiting the next crop if residue is not eliminated on time. Crop residue can be buried or used to make compost that will help improve plant fertilisation. It is not convenient to leave it on the ground since the pest can stay under the stubble waiting for the new plants.

3.6.3.4 Ethological Control

Ethology is the science that studies the behaviour and preferences of insects. Ethological control considers preferences of pest insects. Therefore, this form of control requires specialized, detailed research and deep knowledge of the behaviour of pest insects.

As follows, the most frequently used methods of ethological controls are described:

- **Yellow traps**:
  Wasps, aphids, and whiteflies are attracted by yellow; that is why yellow plastic or cardboard pieces covered in oil are used by ethological control to attract pests. Once attracted, insects get stuck onto the trap due to the oil; unable to escape, their number decreases.
- **Light Traps**:
  Light traps are used to attract nocturnal butterflies and beetles. They consist of a small light bulb with a bowl of water at the base. Insects are attracted by the light and when they fly into it, they fall into the water and drown.
- **Pheromone trap**:
  Pheromones are insect hormones. Pheromone traps attract male insects that are caught either by the water in the base or by an adherent toxic substance. On the other hand, with the dispersion of the hormone in the air, the male insects do not locate females, so the number of eggs immediately decreases. These traps are effective for the first stage of the pest when it has still not spread.

3.6.3.5 Genetic Control

Genetic control consists in making sure that the plant has genes resistant to pests. However, it requires abundant research to determine the genetic characteristics and subsequent research to learn how to incorporate the genes into the population of ornamental plants. This type of control has been used by farmers since time immemorial who have known how to choose resistant plant for farming. Nowadays, genetic engineering develops in laboratories is commonly used to develop transgenics. Therefore, we can distinguish two forms of genetic control:

- **Variety selection**:
  Variety selection is done by farmers for their fields in the countryside and by scientists in experimental fields. Plants have developed different levels of resistance to pests, depending on their adaptation process. Pest resistance can be affected by variety, according to genetic composition, but it can also be influenced by environmental conditions which will enable or prevent plants from showing their resistance in the fields and promote the process of natural selection. Thus, it is essential to choose plants resistant to pests in the same environmental conditions that they are going to withstand. In case of introduced plants, the effect of the environment on them is unknown so it is more convenient to select native plants instead of bringing plants from far away.

- **Transgenics**:
  Transgenic use genes from a different species implanted in the genetic code of the species to be planted in order to increase resistance to pests or introduce new characteristics. The use of transgenics is highly questioned since it introduces new genes which are not part of the genetic selection of the species and could affect the group of genes that conform the population through pollination, affecting wild populations seriously (Martin and Sauerborn, 2006: 182). For this reason, Peru has passed laws that forbid the planting of transgenics for 10 years, until there is more research on the subject. One of the risks of using transgenic crops is that once the more resistant plan species is created, by the natural selection process, pests will also become more resistant and difficult to control.

3.6.3.6 Chemical Control

Even though chemical control leaves residues that may be hazardous to our health, it is the most frequently used form of control. In Peru, many of these chemicals are used in public spaces, unlike Europe where mechanical control is the most frequently used (Ameret al. 211 cit. Wohlgemuth et al. 2019: 284). Based on how long it takes residual effects to vanish, insecticides will be classified according to their level of toxicity. Chemical products use coloured stipes on their labels to show toxicity. In this way: green shows slight toxicity, blue is moderate toxicity, yellow is high toxicity and red is extreme toxicity.

Before applying chemical products, it is necessary to consider some precautions: It is essential to follow instructions on doses and length of time of residual effects for the insecticide. People in charge of applying chemical products are prone to suffer consequences. Therefore, they need to be completely covered with long sleeved shirts as well as long pants and a mask. It is important to apply the product against the wind in order to prevent it from blowing the chemical onto the person that is applying the product.

Choosing when to apply the insecticide is important for the effect it will have on the pest as well as for its residual toxicity. If it is applied during the flowering season it could damage the flowers. If it is applied while harvesting, the length of time for residual toxicity should be considered or crops could damage the consumer of the produce which could be humans, birds or beneficial insects, thus, damaging the trophic chain. For example, if it is applied during the flowering season, it could be absorbed by bees and passed onto the honey damaging quality. Bees are very important for flower pollination and they are currently in danger due to insecticides and climate change which has modified its microhabitat and increased competition with parasitic mites. Without bees, flowers will suffer the consequences and consequently our nourishment will too.

3.7 Nutritional Deficiencies

The most effective way to control pests and disease is ensuring that plants get proper nourishment. If so, the plant will not only be less prone to insect attacks, but it will also be able to recover more quickly from one. As follows we will describe the main deficiencies found in ornamental plants.

Nitrogen deficiency:
Nitrogen deficiency is frequent since it is the element that is required in the highest doses. It ensures the proper growth of the leaves. When there is a deficiency, young leaves are yellowish, the plant is feeble and suffers from dwarfism. Excess of nitrogen causes a delay in the development of flowers and fruit increasing the development of leaves. Furthermore, it can hurt the environment seriously.

Phosphorus Deficiency:
Phosphorus deficiency is frequent in acid soil due to antagonism with other nutrients that compete with phosphorus in the ionic exchange or adhere to it forming compounds in the soil. When a plant has phosphorus deficiency, young leaves turn violet. Furthermore, the number of secondary roots increases while the general development of the root decreases. Without proper roots, flowers and seeds grow poorly. Phosphorus gives flowers a nice colour. With suitable fertilisation using organic matter we ensure the content of phosphorus in the soil.

Potassium deficiency:
Potassium is a macro nutrient that comes together with others; it is often adhered to organic matter. This nutrient is essential for the opening of stomatas and for the plant's ionic balance. Potassium deficiency is shown in the necrosis of the edges of the leaves. If there is a deficiency of potassium when seeds and flowers are forming, these will be poorly developed. Adding organic matter ensure proper nourishment.

Micronutrient deficiency:
A deficiency in micronutrients prevents a plant from forming properly and developing good resistance, making it week and sensitive to pests and disease. There are enough micronutrients in organic matter so to ensure a good source for the plant it should be applied once a year.

Pollution:
Pollution is a problem in the city, especially in areas with a lot of traffic. As a consequence, photosynthesis decreases because stomatas are obstructed by pollutants so there cannot be a proper exchange of gases which stunts plant growth and may cause the premature death of the plant. Plants exposed to pollution have leaves covered in deposits of black pollutants. They lose their colour and show burns between leaf veins. When pollution is very high, the plant becomes more sensitive to pests and disease. In Lima pollution is associated to whiteflies and sooty mould.

3.8 Rule for a Good Garden

Even though each space has special characteristics, there are certain rules for the proper maintenance of plants and green spaces. To have a successful garden, the following should be considered when managing it:

- **Do not drown the plants**:
  It is always important to remember that plants do not only absorb water, but they also need air to breathe. If plants are given too much water, roots rot and the plant dies. Therefore, it is vital to control watering.
- **Let plants rest**:
  After flowering and bearing fruit, a plant is tired since it has used up all its nutrients for production. Therefore, so that the plant can again give big and beautiful flower, the plant needs to rest for a period. If it is perennial and semi-perennial, at least 3 months is required as long as it is fertilised after the harvest. If seeds are not going to be used, flowers can be cut after blooming, then the plant will not use nutrients for bearing fruit and will recover faster. It is also useful to remove old flowers and leaves to prevent pests and disease.
- **Replace old plants**:
  As plants get older, they become more sensitive to pests and disease and require more care. Therefore, age should always be considered so that the decision to replace the plants is timely. Otherwise, the green space will require more care. Furthermore, it will be more costly, and results will be less pleasant.
- **Make a good selection of plants**:
  A garden needs a selection of plants that require the same conditions of water, light and soil. Otherwise, maintenance is too difficult, and some plants may die because they do not have the right conditions to grow. It is important to know the origin of the plant. This is not so difficult since plants with equal external characteristics usually come from the same place. But in case of doubt, it is advisable to search for its origin.
- **Group plants adequately**:
  Plant grouping can be difficult especially considering that plants from the same species are attacked by the same pests. Moreover, if the plant beds are big it will not be easy to eliminate pests by means of biological control. Therefore, we should not design beds that are too big choosing plants of only one species. It is advisable to use different species, forming groups, strips or circles. Results are better when plants come from different families. By doing so, we make sure pest

insects are different for each species selected and we diminish the risk of pests.

- **Give each plant enough space**:
  So that plants can grow and feed, there should be space between one another taking into account the fourth dimension in order to prevent competition of plants of the same species as well as plants of different species.

References

Endlicher W, Kraas F, Krajewski Ch (2012) Einführung in die Stadökologie. UTB, Stuttgart, p 272

Francis R, Chadwick M (2013) Urban ecosystems: Understanding the Human Environment. Routledge, USA, p 220

Kluckert E (2000) Grandes jardines de Europa: desde la antigüedad hasta nuestros días. Könneman, Colonia, p 496

Martin K, Sauerborn J (2006) Agrarökolgie. Ulmer UTB, Stuttgart, p 297

Muñoz F (1979) Como puede Ud. diseñar su propio jardín, Manual práctico de diseño. Universidad Nacional Agraria La Molina. Lima

Real Academia Española (2019) Diccionario de la Real academia española. https://dle.rae.es/?id=MMXffef. Revised: 3.01.2020.

Sademann A, Kilimann S (eds) (2017) Berlin dr grüne Stadtausflug. Ed: via reise tours verlag. Berlin, p 168

Smith T, Smith RL (2007) Ecología, 6th edn. Pearson Educación S A. Madrid, p 776

Swoczyna T, Borowski J, Latoja P (2017) Trees and shrubs for urban plantings: introduced or native species. In: Congres presentation: problems of landscape protection and management in XXI century. Organized by Warsaw University of Life Sciences, Polski klub ekolgiszny

Vercelloni M, Vercelloni V (2010) Geschichte der Gartenkultur von der Antike bis heute. WBG, Darmstad, p 275

Wohlgemuth T, Jentsch A, Seidl R (eds) (2019) Störungsökologie. Utb Haupt Verlag, Gernany, p 396

Peruvian Gardens

4

Abstract

This chapter focuses on Peruvian landscape in Lima. It describes the development of landscape concepts and their implementation in Peru considering their history. It describes the current situation of landscape in Peru distinguishing between parks on the coast, the highlands and the jungle, taking examples and describing the most representative parks and gardens of each region with their own climatic characteristics and cultural diversity. This chapter also describes the development and adoption of urban bio-orchard concepts in Peru, specially in Lima. It proposes new options for the development of green areas in the twenty-first century considering botanical, agronomic, socio-cultural and landscaping aspects. The chapter makes possible, with practical examples of the implementations of new green areas in Peru to make Lima and other cities of Peru livable and sustainable, and contribute in this way to the Sustainable Development Goal 11. The section that presents the Bio-orchards provides practical examples to ensure the food security, healthy and accessible food and contributes in this way to the Sustainable Development Goal 2 and 12. Finally, it describes and analyzes the main botanical species used in the history of the gardens in Lima.

Keyword

Peru · History · Coast · Highlands · Jungle · Bio-orchards

This chapter focuses on Peruvian landscape in Lima. It describes the development of landscape concepts and their implementation in Peru considering their history. It describes the current situation of landscape in Peru distinguishing between parks on the coast, the highlands and the jungle, taking examples and describing the most representative parks and gardens of each region with their own climatic characteristics and cultural diversity. This chapter also describes the development and adoption of urban bio-orchard concepts in Peru, specially in Lima. It proposes new options for the development of green areas in the twenty-first century considering botanical, agronomic, socio-cultural and landscaping aspects. The chapter makes possible, with practical examples of the implementations of new green areas in Peru to make Lima and other cities of Peru livable and sustainable, and contribute in this way to the Sustainable Development Goal 11. The section that presents the Bio-orchards provides practical examples to ensure the food security, healthy and accessible food and contributes in this way to the Sustainable Development Goal 2 and 12.

A. Sabogal, *Urban Ecology*, Sustainable Development Goals Series,
https://doi.org/10.1007/978-3-030-69905-5_4

Finally, it describes and analyzes the main botanical species used in the history of the gardens in Lima.

4.1 History of Gardens in Lima

The idea of a park was brought to Peru during the Spanish conquest. During the Inca Empire, public spaces were similar to what we know today as squares, places where people gather for assemblies, with no green spaces but with view to beautiful landscape. This view was important to Incan design not only for its beauty but also to ensure dominance over space. For example, in Machupichu, most windows are directed to outdoor landscape. This marks a substantial difference with the Arab culture and the architecture of cloisters brought by the Spanish that direct doors and windows indoors which is distinctive in the traditional gardens in Lima.

Parks and gardens in Lima are rooted in the Arab gardens which were brought to us by the Spanish, thus, we often find squares and orchards in them. Like gardens in the south of Spain, Lima's gardens are placed in spaces where water is scarce and basically ornamental with no large extensions of green. In this sense, Lima is a problematic space for the development of parks. Park planning should always consider that there is not enough water. Therefore, the Arab garden is the most adequate legacy for this climate.

The first gardens built by the Spanish during the colonies were indoors, in convents or orchards in houses as well as walkways covered in tile with sculptures and water fountains but closed to the public. It was not until the twentieth century with the creation of the "Parque de la Exposición", in the outskirts of Lima city that parks are granted a socio-cultural role for family walks and recreation linked to the idea of health and usefulness given to the parks nowadays.

When strolling around most parks in Lima, we can see that plants are pruned evenly, expressing the dominance of man over nature. The concept of park is still understood as a place with perfect symmetry with no space for disorderly plants and everything is in line; nature is held back, tamed by man; a park serves man and in it man has a power over nature modelling it at taste. Furthermore, under this view, a park is only important when it has a building, a monument, a bandstand; These buildings are what matters in the park. Only in this way do we feel liberated from social problems and away from the dangers in nature, still nearby. On the other hand, parks with free nature, in which branches change forms to surprise walkers and enable birds to nest and beneficial fauna to develop, can only prosper with moderate pruning that allows development. Drastic pruning during blooming season will prevent the development of flowers and seeds so trees will not be able to complete their vital cycle (see chapter 3), However, a park develops another dimension when plants reach adulthood and trees have a history.

Throughout the history of Lima several orchards were built and then came the parks. One of the first orchards was the one in San Lazaro's church, located on Jiron Trujillo in Rimac district, which was the first leprosarium in Lima, built 450 years ago in 1563 just a few years after Lima was founded (see Fig. 4.1). Even though it is currently located in the middle of the city, at the time it was in the outskirts, behind the government palace, on the other side of Rimac river where the ill were isolated from the rest of the population. In this place the ill had their own orchard to supply them with herbs and vegetables. It is said that the fig tree in the orchard was planted in the time of Pizarro. Nowadays, even though the small church is still kept for its cultural beauty but only with one of its towers, the orchard is destroyed and abandoned and given no value by the church or the population in spite of its great cultural and historical value. Additional to important paintings, the church also has the processional cross of Amancaes that was carried in processions to the hills where the amancaes (*Ismene Amancaes*) bloomed.

The "Alameda de los Descalzos", also located in Rimac district, built by viceroy Juan Mendoza y Luna between 1609 and 1611 reminisces an Arab park so typical of Moorish Spain. This "alameda", with water fountains and beautiful

Fig. 4.1 San Lazaro church. *Author* Ana Sabogal

sculptures typical of walkways for the viceroy and wealthy people of the time, hardly had green spaces, only a little grass in the side berms and some marble flower pots that most certainly had flowers at the time (see Fig. 4.2). The design is that of a French park intended for walks, musical performances or public events. This "alameda" is currently located in the middle of Lima adjacent to San Cristobal Hill which is now overpopulated with housing in very poor conditions. City growth in Lima has caused this park to be abandoned in an area with very little financial resources in the Rimac district. However, some of the sculptures are still intact, and the beauty of the marble pots can still be appreciated.

Most of these historic parks from viceroyal times have changed enormously due to city growth. However, they still have a very significant combination of green areas and art, important sculptures distinctive of these spaces. It takes us to a time that parks were intended for strolls by wealthy Spanish and Indians in Lima.

Among the gardens, the one in the "Quinta Heeren" stands out, located between Junín and Maynas avenues in Barrios Altos, a neighbourhood in old Lima centre. It was built in 1880 (Vernal 2000). Land was inherited by the daughter of President José Pardo y Barreda, Carmen Ignacia Barreda y Osma, married to a German citizen Oscar Herren. Both had the Quinta Herren built and named it after its owner. Before marriage, Oscar Herren had been the first honorary General Consul of Peru in Japan, so he knew the Japanese culture (PUCP 2016). The Quinta Heeren had a beautiful European style garden for the residence of diplomats (Municipalidad de Lima and OEI 1998). The big 4-hectare piece of land (Guía Expreso 1998), allowed for an extraordinary garden. It was an architectonic complex that included houses, a garden and a small square of Austrian neoclassic *Biedermeier* style, in fashion at the time (Guía Expreso 1998). The small square was decorated with beautiful marble structures. The garden included a botanical garden and a zoo with exotic animal like a giraffe and an elephant (Guía Expreso 1998).

The garden was designed and built by garden designer Tatsugoro Matsumoto (Municipalidad de Lima and OEI 1998), who included several exotic species from Japanese gardens. Tatsugoro Matsumoto was well known at the time. After the Quinta was finished, he travelled to Mexico where he is also known for his garden designs. Today the garden of the Quinta Heeren still has a lot of exotic species. Lima's first tennis courts were built there as well as the equestrian tracks (Guía Expreso 1998) since both tennis and horseback riding were sports enjoyed by diplomats. Several embassies were lodged in the Quinta (Guías Expreso 1998).

In 1873 the governments of Peru and Japanese signed the treaty of Peace, Friendship and Commerce, initially negotiated by Oscar Herren. This treaty brought the Japanese population to work on the rubber plantations (Lausent-Herrera 1991). The decadence of the Quinta Herren began when a prosperous Japanese merchant, Seikuna Kitsutani, who was living in the Quinta, lost his fortune after an accident with two of his shipment transporting workers (culíes). He committed *haraquiri* in his house of the Quinta right after (Municipalidad de Lima and OEI

Fig. 4.2 Alameda de los descalzos. *Author* Ana Sabogal

1998). The truth is that slowly the Quinta became part of Lima centre which was getting more densely populated and was being abandoned by the wealthy who were moving to less populated districts. Currently the Quinta is completely abandoned. During the twentieth century it suffered many changes first with the earthquake in the 40s that damaged its construction seriously and then with the invasions in the 50s, after which much of the land was rented for agriculture (Municipalidad de Lima and OEI 1998). The Quinta Heeren was declared a National Cultural Patrimony by the "Instituto Nacional de Cultura del Perú" (INC) in 1972 for its significant architectonic, landscape, cultural and artistic value (see Fig. 4.3). Currently, the Ministerio de Cultura (Culture Ministry) has a plan to restore the Quinta due to its historic and cultural value and as a cultural monument to humanity (El Comercio 2019).

The central plaza or Plaza de Armas is a characteristic Spanish construction known as Damero de Pizarro (checkboard of Pizarro) in there the Church, the government and the municipality enclose the square (see Fig. 4.4).

The "*Parque de la Exposición*" was created in the outskirts of Lima at the end of the nineteenth century when President Balta decides to tear down the city's south wall to host the 1872 Great International Exhibition. This park was then

Fig. 4.3 Quinta Herren. *Author* Ana Sabogal

Fig. 4.4 Central plaza, Lima. *Author* Ana Sabogal

bigger than it is now, containing the "*Palacio de la Exposición*". Within the doors of the palace is the Lima Art Museum as well as zones that do not belong to the park anymore such as the Italian Museum, Civic Centre and the National Stadium. The park was designed by two renown architects of the time, Manuel Anastasio Fuentes and Antonio Leonardi. It also included a zoo. For the celebration of Lima's Independence Centenary in 1921, several new pavilions were built in the parks. Artists of the time were called upon such as José Sabogal who built a "*costumbrismo*" fountain Furthermore, embassies donated monuments that today are important features in the park (see Fig. 4.5). With time, the park became a central part of the city immersed in Lima centre. In 1970, due to Lima's enormous growth and all the prevailing social problems, the fence that we can see nowadays was built around the park. In 1973, one hundred years after the signature of the treaty of Peace, Friendship and Commerce, the Japanese colony donated the Japanese park that became part of the "*Parque de la Exposición*" (see Fig. 4.6).

Recently the "*Parque de la Exposición*" has been renamed the "Gran Parque de Lima", and includes historical monuments, a small theatre and a great open-air theatre which is currently used for shows that are attended by a big audience with space for 4200 spectators. It is a place where popular spontaneous theatre is possible as well as the sale of popular food and colourful balloons. There is also an area for a botanical garden and all car circulation has been eliminated, thus, it has the characteristics of a zonal park. In general, the new architecture of the park has a lot of cement and little space for plants. However, it still has some species of old trees with history. It is visited by all the inhabitants of Lima and it is one of the few options for the inhabitants of this area. It reminds us of the concept of a Spanish park with "alamedas" and squares suitable for shows.

4.1.1 Gardens in Lima

Lima was founded in 1535, in a strategic location in front of the sea and in the valley of Rimac river. Its proximity to Callao port assured control of commerce with the Spanish crown and its proximity to the valley made it possible to have a lot of green spaces and agricultural fields around Lima ensuring food for the city. That is why Lima is known as the "City of Gardens". Currently, Lima has grown to be 2672 Km^2 big. From the ecological point of view, Lima covers two axles. On one side, the Rimac river; first axle over which the city was founded and covers the width of the city. On the other hand, the axle along the Pacific Ocean coast covering the city from north to south, which currently comprises two river basins: Chillon river in the north and Lurin river in the south. Lima currently covers 3 valleys and great part of the land is desertic.

Since the 1950s the number of people living in the cities has increased considerably and Lima is no exception. It is estimated that in the fifties, 28.8% of the world population lived in cities, whereas, by the end of the twentieth century this figure grew to 46.4% and by 2007 the figure reached 50% of the world population (Endlicher

Fig. 4.5 Parque de la Exposición, pavillon. *Author* Ana Sabogal

Fig. 4.6 Parque de la Exposición, Japanese garden. *Author* Ana Sabogal

et al. 2012). Throughout history there has always been migration to Lima, but this was especially so in the twentieth century. In the 1940s the population in Lima was only 16,6% of the inhabitants in Peru with slightly more than 600,000 inhabitants (INEI 2008). President Odria, dictator known for implementing a social system, transformed Lima into an urban centre with medical facilities and housing in the 50s. In the 60s, the number of inhabitants in Lima was 1,7 million, and in the 80s it reached 4,5 million (INEI 2008). The agrarian reform in 1969 and terrorism in the 80 and 90s were crucial in the rampant growth of the city, causing a strip of poverty to develop around Lima. As a result, in 1993 Lima's population was 28.7% of the total population in Peru. By 2007 it reached 30,9% (INEI 2008). Currently, Lima has around 10 million inhabitants which is 30% of the country's population (INEI 2015). The main reason for migration is centralization in the capital which causes lack of opportunities outside the city. Here is where hospitals, schools and universities concentrate. Whereas in the country in 2011 the level of poverty was 27.8% of the population and extreme poverty was 6.3%, in the city of Lima the level of poverty was 15.6% of the population, and extreme poverty was 0.5% (INEI 2013). We can see a great difference between the population in the countryside and in the city, which is the main reason for migration.

Urban growth in Lima has been centripetal, from the historic centre to the borders. The historic centre has been abandoned slowly by the wealthy who are moving to the periphery. However, there are still many buildings in terrible conditions. The periphery has developed with a lot of inequality. On one hand, slums have developed to the south and north and big houses have been built by the wealthy in the east and west. Even though Lima has changed in the last years and there are more buildings than before, with exception of some districts, it is a city with low constructions, and it is no longer a garden city. It has few green areas especially when we consider green spaces in relation with number of inhabitants.

The concept of *Garden city* is born in England and it is postulated by Ebenezer Howard in 1898, who plans cities with spaces for agriculture in strips surrounding the industrial cities and intermediate cities interconnected with agricultural spaces. The city is developed in concentric circles with parks and groves interconnecting spaces and the industrial space surrounding it in an external circle (Endlicher et al. 2012). The concept of **Garden** used in this research is based on the definition given by the "Diccionario de la Real Academia Española"(Dictionary of the Spanish Royal Academy): "Land where plants are grown with an ornamental purpose", distinguishing it from the term **garden city** defined by the Royal Academy of Spanish as: "an urban group formed by single-family houses each one with a garden" (Diccionario de la Real Academia Española 2019).

In cities there is a relationship between economic level and the amount and proximity to green spaces as well as species diversity (Francis and Chadwick 2013: 70). Green spaces in the city are considered a luxury. There is a relationship between purchasing power and public spaces; whereas high class areas have few public spaces, middle class areas where people do not have the means for private green spaces but pay taxes, have more public spaces per inhabitants (Sabogal et al. 2019). These spaces are also used by inhabitants from marginal zones who go to public spaces in neighbouring districts (Sabogal et al. 2019). City statistics do not distinguish between private and public urban green spaces so this difference of distribution and access to public spaces cannot be perceived (Sabogal et al. 2019). Districts with high purchasing power have enough green spaces; such is the case of San Isidro with 19.92 m^2/inhab., Miraflores with 13.76 m^2/inhab. and Jesús María with 9.02 m^2/inhab.; whereas other low-income districts have very little, like Villa María Triunfo with only 0.34 m^2/inhab., Breña with 0.37 m^2/inhab., San Juan de Lurigancho with 0,61 m^2/inhab (Lima como vamos 2014). This, together with pollution caused by a large and deficient vehicle fleet and chaotic public transportation increases the risk of respiratory illnesses and diarrhea which among the inhabitants in Lima was 6% in 2009 (Liebenthal and Salvenini 2011).

4.2 Peruvian Garden

Spanish bring to Peru the heritage of Andalusian parks, that at the same time inherit extensively the characteristics of Arab parks. It is a park where water is scarce and very valuable, so there are no great extensions of grass. We have inherited spaces with tiles and fountains of water. The first gardens were indoor gardens in convents or orchards in houses. Gardens on the coast follow this pattern, especially in Lima, as the centre of the viceroyalty.

4.2.1 Gardens on the Coast

Since the end of the twentieth century and especially in the twenty-first century in Lima, once the economy improved in Peru, the development of green areas began as well as the debate about these areas. This coincided with the increase of the population in the cities and the depopulation of the countryside. This description will use as example the parks of the coast. Among the Parks in Lima we will describe the "Alameda de la Juventud" and the Costa Verde corridor. We will describe is the zonal park of Chimbote in the city of Chimbote located to the north of Lima (see Sect. 4.5.2).

The park of the "*Alameda de Juventud*" is located in the south of Lima in Villa El Salvador district along Juan Velasco avenue, in the crossing of Los Alamos and Pastor Sevilla avenues. It is a wide avenue with an unpaved central berm, which was filled with domestic garbage before the park was built. Originally the land was just a sandy lot adjacent to a hill called "Lomo de Corvina" which connects to the Panamericana Sur highway and then to the beach explaining why it is so sandy. What is interesting about this terrain is the steep slope it is on. Villa El Salvador is one of the most populated districts in Lima City whose main economic activities are industry, commerce and in the periphery of the district, urban agriculture of basic products.

The "Alameda de la Juventud" was inaugurated on March 25, 2001. It was designed by the non-governmental organization "*Centro Estudios y Promoción del Desarrollo* (DESCO)", together with the "*Centro de Investigación de Proyectos Urbano Regionales* (CIPUR)" and the Villa El Salvador Municipality. This project is born with the intention to create a green area using the water from the water treatment plant. The space was designed as a big "alameda" with small squares, spaces for children and sports as well as chess tables for competitions. It is an "alameda" along the avenue where instead of poplars there are native trees. This was one of the first works done in the park which includes essential ecological aspects such as water recycling, selection of native plants, plant nurseries and even aspects of inclusive design, among others.

The development of the project of the "Alameda de la Juventud" includes the following:

- A primary aerobic water treatment plant that enables access to water for irrigation for the park. This plant was not easily accepted by the population because the water released bad smells and was not potable, being the people in the area accustomed to drinking water from hoses in the parks when thirsty. Water with only primary treatment should not be drunk or used for washing hands or face.
- An irrigation system with sprinklers had to be adapted for water with primary treatment since the snoozes in the system would clog. Therefore, wider snoozes had to be considered as well as more frequent maintenance.
- A pedestrian recreational zone included chess tables, games for children, pergola, benches and urban furniture.
- Area for a financially self-sufficient ornamental plant nursery to supply the "Alameda" with plants.

  - An educational area composed by:
  - A collection of plants classified based on the most significant ecological zones in Peru. Among the tree species planted for the coast we can find tara (*Caesalpinia tinctoria*), palo verde (*Parkinsonia aculeata*), carob tree (*Prosopis juliflora*), among others.
  - An area of medicinal plants with the ones most often used by the population such as mint (*Mentha spicata*), oregano (*Origanum vulgare*), cedar (Aloysia citrodora) or lemongrass (*Cymbopogon citratus*), among other.
  - An area of exotic or introduced species such as the umbrella tree (*Schefflera actinophylla*), the croton (*Codiaeum variegatum*) or the poinciana (*Delonix regia*).
  - An area of xerophilous plants suitable for arid areas such as: aloe (*Aloe vera*) and Calanchoe (*Kalanchoe flammea*), among others.

- An aerobic water treatment plant designed with a circuit for environmental education that can be visited by the population.

- The "alameda" was formed with homogeneous sectors to give it the character of an "alameda". The first sector was planted with the bottlebrush trees (*Callistemon citrinus*), which adapts to low humidity and is translucent so that it connects spaces of the park with areas for pedestrians in the central and lateral berms. The second sector was planted with bottle palm (*Hyophorbe lagenicaulis*) adapted to the desertic areas and intense wind. The third sector was planted with the Peruvian pepper tree (*Schinus molle*), a native plant common in the ecosystem of the Rimac river basin.
- Low plants were chosen based on their low water requirements and trampling tolerance, giving preference to Peruvian plants. Flower plants selected had to be perennial to reduce maintenance and costs. The species chosen were: lantanas (*Lantana camara, Lantana montevidensis*), verbena (*Verbena peruviana*), ichu (*Festuca spp.*), Chinese carnation (*Mesembryanthemum spectabile*), among others.
- Pergolas were covered with Bougainvillea (*Bouganvillea spectabilis*) of a variety of colours that come from the north of Peru. In this way, the different sectors of the "alameda" can be distinguished, but at the same time they maintain the uniformity of the space but break the monotony with this variety of colours.
- Low species considered were the xerophilous such as the Mexican cactus (*Euphorbia candelabrum*), the calanchoe (*Calanchoe flamea*), among others.
- The urban furniture such as the fence around the plant nursery, sheds, pergolas, etc. were built with local Guayaquil cane. The design of urban furniture was conceived with local material.

With the choice of species, the design intends to highlight the playful liberty of nature in contrast with the symmetry of the architectonic design which is counterposed with the mobility of the plants chosen for the "*alameda*". For this reason, the plants are mobile and translucent. This also allow all spaces and the surroundings to integrate. When selecting the plants, blooming was also considered in order to have a colourful "alameda" all year around. Other parameters considered for plant selection in order to ensure proper development and beauty of the park were that plants needed to be easy to care with no need for pruning, adapted to climate, resistant to water fluctuations with a variety of blooming seasons. Additionally, plants required individual characteristics as well as value for the population.

The "*Alameda de la Juventud*" is the first initiative with active participation of the local population for its implementation. That is why great importance was given to local medical plants used by the local population who mainly comes from the countryside. It is important to point out that the population in Villa El Salvador arrived in the 1970s during the migration from the countryside to Lima city. Most of the population in Villa El Salvador come from the highlands of Ayacucho.

The design also had to consider practical criterions like the possibility of finding the plants easily in the plant nurseries in Lima and the specific characteristics of the species like toxicity and root invasiveness (see Fig. 4.7).

Another interesting proposal developed in the last years is the chain of parks along the coastline from Barranco district located in the south of Lima city to San Miguel district located in the north of the city, forming an ecological corridor with few interruptions, enabling the population to walk for several kilometres. This ecological corridor has been well received by the population especially in Miraflores district which is the portion with the most parks (see Fig. 4.8). Nowadays, on weekends many people stroll along the corridor, large families and teenagers, many on picnic or practicing a sport. Many of the people come from districts with few green areas. In this area there are also organized cultural activities as well as free ones like yoga, juggling, theatre, tight rope, paragliding, among others, giving Lima a new life and integrating generations.

Fig. 4.7 Design and implementation of the "Alameda de la Juventud". *Author* Ana Sabogal

Fig. 4.8 Costa Verde ecological corridor, Miraflores boardwalk. *Author* Ana Sabogal

4.2.2 Garden in the Highlands

The concept of garden in the Peruvian Highlands is one where the city stands out against its surrounding with its edifications encompassed by the countryside. Even though this is changing slowly, there is a great difference between the natural surroundings and the city. Another characteristic of this space is the altitude which causes low temperature and limits the species that can adapt to the climate of the highlands. In this context the parks fulfil the role of being a place where people meet on holidays and celebrations. The inhabitants perceive parks in relationship to the security provided by constructed spaces and as a place to meet people in markets and squares. In traditional cities like Cusco there is a Spanish legacy in the perception of parks as well as an Inca legacy. Both perceptions see the square as a place to meet and celebrate festivities.

Unlike cities on the coast, there is no search for natural spaces.

It is important for the inhabitants of the highlands to have a view to the landscape. The view of a space from the mountain is common in Inca design. This form of design looks for the domination of the landscape and the view of the valley from the top and can be appreciated in the most important Inca edifications like the Inca city of Machu Pichu and the Inca archeologic complexes of temples of Pisac, both with views from the mountain to the gorge.

In the highlands of Huarochiri, in Lima, we can find a small town called San Jerónimo de Surco located on the "carretera central" at 2008 m above sea level, with approximately 1400 inhabitants. This town was founded almost 100 years ago. Its structure is traditional which means that the most important public space is the square. This square is traditional, surrounded by the main public buildings which are the municipality, the school, the police station and the church. The square is where parades, festivities and other public events take place. Another important green space, like in many towns in the highlands, is the cemetery. It has a lot of flowers and trees and it is visited by many inhabitants not only on festive days. Finally, the only green space designed as such is a small park located at the town entrance with two cement benches, each under the shade of a ficus benjamina (*Ficus benjamina*) under which the space has been outlined with a cement border and grass (see Fig. 4.9). We were surprised by the view to the gorge with the Rimac river at the bottom. It is an open view with the highway full of trucks going to Lima city loaded with produce. We can see that the perception of public space depends on a context and culture. It is interesting to also observe that this perception is linked not so much to a need than to a longing and to life ideals that change with time and context, leading us to their cultural heritage.

Another interesting route in the highlands is the one from Lima to Canta, along the Chillon river basin. In this space, encased by the basin and flanked by an enormous gorge of hundreds of meters, just like many other roads in the highlands of Peru, right before reaching Canta, you meet with one of the few brick constructions, a house built on the edge of a cliff, inhabited by only one elderly woman native of the area. Her sons and daughters migrated to United States, but she preferred to stay in her place of birth. They had this house built for her following her instructions at the edge of the cliff so that she could contemplate nature. We can see that landscape and spaces are representations and constructions of perception. Therefore, it is difficult to talk about what is appropriate. This will depend on our cultural constructions, personal experiences, smells, sounds and the taste of childhood.

4.2.3 Garden in the Jungle

The Amazon is characterized by the abundance of vegetation. Thus, we always associate it with natural spaces. However, nowadays with the development and expansion of the cities great part of the population live in them. They are emerging cities shaped by commerce. In most of them the topic of green spaces has yet to be proposed. Even though many of them like Iquitos and Moyobamba were founded centuries ago, their green areas are limited to the main square located in the city centre. Moyobamba, the first city founded in the jungle in 1540, was developed to promote the development of other cities in the Amazones. Iquitos was founded time after in 1740 to expand the catholic church lead by the Jesuit mission. All these cities developed slowly. Iquitos went through economic prosperity during the time of the rubber boom. Nowadays, we still find beautiful mansions with tile brought from Portugal. Finally, Pucallpa city, founded in 1883 due to the expansion of the rubber boom, expanded in the twentieth century with the exploitation of timber and the implementation of the highway.

In the cities of the Peruvian jungle rivers are an important commercial axel since they are the primary communication route. For this reason, additionally to the main square, another essential public space is the boardwalk along the river that

Fig. 4.9 San Gerónimo de Surco park. *Author* Ana Sabogal

is the main ecological corridor and green area in the city. Along the riverbank, we can find the market visited by the residents of the area. Furthermore, the river is where the produce arrives. We can find markets on riverbanks of the cities of Iquitos, Pucallpa and Moyobamba with boardwalks (see Fig. 4.10).

Fig. 4.10 Boardwalk on the riverbank of Itaya river, tributary of Amazonas river, city of Iquitos. *Photo* Ana Sabogal

4.3 Bio-Orchards

In these times, the population has distanced from natural ecosystems in a society where many children have never seen the animals their food comes from, nor have they seen how the plants they eat grow. Therefore, the population has developed new forms of approach to nature. This is possible due to new agronomic and ecologic technics incorporating all the components of the trophic chain to the design.

However, the idea of bio-orchards is not new. Since the nineteenth century, the need for green areas is acknowledged and doctor Leberecht Migge (1881–1935) who describes the relationship between green areas and health expresses this concept in bio-orchards. The so called *Schrebergarten*, named after German doctor Moritz Schreber, are a response to the industrialization of the cities. These small gardens were grown in the outskirts of the cities in the mid nineteen century. Ernest Hauschild, school principle, together with the parents' association of the school and the students built the first *Schrebergarten* for the entertainment of the children whose parents worked at the factories (Endlicher et al. 2012: 198–199). The *Schrebergarten* were composed by bio-orchards in the outskirts of the cities, where workers could spend their weekends growing vegetables and fruit while coexisting with nature (Vercelloni and Vercelloni 2010: 223). Subsequently, Le Corbusier in his design "*Plano Voisin*, to restore the city of Paris in 1925, included the implementation of bio-orchards as part of the green areas.

Currently Germany has 1,24 million small gardens (BMVBS 2008 cit. Endlicher et al. 2012: 1997). According to German regulations (Bundeskleingartengesetz 1983), small gardens should not be bigger than 400 m^2, whereas constructions within them should not be bigger than $24m^2$. Gardens should consider ecological precepts (Endlicher et al. 2012: 198). The population that uses these gardens are very involved with ecology to such an extent that 96% of the users of these small gardens make their own compost (Breuste 1996 cit. Endlicher et al. 2012: 196–197). The small bio-orchards are only used during warm months and are closed in winter. The population uses the gardens on weekend for entertainment and they use fruit and vegetables collected to cook.

Currently the Small Gardens are either Community Gardens or Bio-orchards managed by groups of citizens pursuant to municipal regulations of citizens. They can be found in Germany, Sweden and Australia where they are greatly accepted and are often part of spatial design. In Toronto there are about 100 community gardens, some of them within public parks. (Endlicher et al. 2012: 201). It is estimated that there are more than 67,000 small *Schrebergarte* in Berlin, which means that around 2% of the population manages one of these small gardens.

But all this goes beyond the growing of vegetables; It is to a great extent a proposal for a different style of life. Worldwide, in the last decades the subject of urban agriculture has been part of many projects, many of them implemented in South America. The debate about cheap food available for everyone has become an issue related to quality of life of local populations. Quality certifications of food, the great amount of low-quality chemical products that pose health risks are also part of the debate as well as the worries of city dwellers. Additionally, there is a discussion about the fair price of product that enables agricultures to live in dignity. We cannot forget Climate Change which will have the greatest impact on cities. Finally, the movement called *slow food* has questioned and proposed a discussion regarding the quality of fast-passed life led by city dwellers. In this sense, we are talking about a social and political movement. (Pollan 2010: 342).

Among other movement facing the described scenario, a current landscaping movement, represented by landscapers Batlle, Roig and Gali, among others, proposes the introduction of cattle into parks as part o f a natural control in the same way as in English gardens (Zabalbeascoa 2017). This allows the natural incorporation of organic matter and the development of the flora and fauna of the soil and it also approximates city inhabitants to the countryside. This concept could be perceived at the beginning of the twentieth century when Fritz Schumacher, who won the bid to design the People's Park in Hamburg, included a stable in his design where there would be a production of milk, cheese and butter, concept which is still a romantic design (Vercelloni and Vercelloni 2010: 222). In this way, ecological concepts and principals are incorporated to park design.

Peri-urban architecture plays an important role in the city, given that 17.5% of inhabitants live in extreme poverty (1.5 million people) (FAO 2014: 62). Examples are found in the peripheral districts of Carabayllo, Puente Piedra, Pachacamac, Lurín, Lurigancho, Chosica and Ate Vitarte, where vegetables, fruit, corn, fodder species, and ornamental plants are cultivated. Residents use these crops for self-consumption or as a means of income as day-laborers or growers on rented land (FAO 2014: 62). Fast growth of Lima City has caused the absorption of important agricultural areas. Slowly, old farms like Santiago de Surco or Boca Negra have become an important part of the city, However, urban and peri-urban agriculture still cover 600 ha. These are small spaces, often located in marginal areas at the edge of the city. We can find pig farms managed in precarious conditions (FAO 2014: 62). Urban orchards are a different reality. They are very small, about 4 m^2 and can be found in Villa El Salvador, Surco and Chorrillos (FAO 2014: 63). They are bio- orchard that do not use chemical products and are watered with drinking water (FAO 2014: 63). Most of the produce grown is for self-consumption, yet, some is sold in farm markets and weekly markets located in high income districts of Lima city.

Some districts in Lima's periphery such as Villa María del Triunfo, Ventanilla, Lurigancho, Chosica and Villa El Salvador have incorporated in management and organization programs agricultural aspects, and the municipality of Lima City has approved a regulation for the promotion of urban agriculture, incorporating spaces for bio-orchards in public parks and other public spaces (FAO 2014: 64). As of 2014, the municipal program *Mi Huerta* benefited 23,000 people—including school children, families, and community members—via 1,000 urban gardens spanning twelve hectares across several districts (Lima Cómo Vamos 2015). The vegetable program of the Universidad Nacional Agraria has promoted the process since the 80 s (see Fig. 4.11).

A new market for agricultural products free from chemical products has developed in Lima in the last decade. Furthermore, a commercialization system to sell products from fair market is being promoted. In Lima city, daily there are more people aware of the effects of pollution and committed to protecting the environment. Lima is not the only city that has suffered from the effects of growth and migration. Many other cities, such as Piura city, have grown and in them we can find many small bio-orchards built by migrant populations.

Fig. 4.11 Bio-orchard in Villa María del Triunfo, Lima. *Author* Ana Sabogal

In the jungle it is difficult to grow introduced vegetables due to high temperatures that to not allow the growth of the most traditional vegetables brought from Europe or Asia. Additionally, high temperatures make it a health risk to consume fresh vegetables. However, the use of herbaceous plants as well as vines or semi-woody plants with a high content of micronutrients for seasoning is part of the local diet but there are only a few studies on the matter.

Inkaterra's experimental station, located in the south of Peru in Puerto Maldonado city, Madre de Dios department, is pioneer in the subject. It has been doing research on how to rescue species for bio orchards considering that the ecosystems in the jungle have many trees and very few open spaces. Therefore, mixed systems are suitable for bio-orchards grown under these conditions. It is important to include vines like passion fruit (*Passiflora edulis*) or granadilla (*Passiflora ligularis*) and bushes like the spiked pepper (*Piper aduncum*) or the banana (*Musa x paradisiaca*), as well as mint (*Mentha piperita*) or hierba buena (*Mentha spicata*) to ensure a balanced diet. In order to rescue traditions while cities continue their growth, it is important to include the migrant local population from the countryside and make them value these bio-orchards so common in the homes in the jungle and in the high jungle, which should include in the orchards low plants as well as vines and bushes and even fruit trees, thus, design a new concept of orchard suitable for the jungle. Plants should grow almost spontaneously and should be essential for people's diet as well as for the development and revaluation of their culture in the cities developing in the jungle.

4.4 Gardens in Lima in the Twenty-First Century

History of the design of green areas can be divided in two stages; in the first stage, the forest or agronomic engineers were the ones that designed parks and in the second stage the architects and space planners designed gardens and public spaces. The twenty-first century tries to unite these two visions. The landscaping proposal for gardens in Lima for the twenty-first century should propose not only how to rediscover space ecology but also how to get along with nature. For this, we should determine which are the natural spaces in modern Lima and what is the history and nature of space. In this sense, it is essential to rescue Rimac river and integrate it into the landscape as a historical and ecological space that should interconnect the city since it goes through it as a spatial and historical artery. In this sense and picking up Jean-Claude Forestier ideas (Sect. 1.4), ecological aspects should be integrated as collective spaces. Therefore, and based on Frank LLoyd Wright's proposal (Sect. 1.4) part of this revalorization depends on space being integrated to the design.

The twenty-first century is definitely about going back to nature and ecology. Rediscovering the essence of the landscape. Retaking the concept of returning nature to space, there should be a bridge between wild nature and the city as well as its history. This concept is recreated in the park of the Paris National Library designed by Gaëlle Larriot-Prevost between 1989 and 1995. Here, the idea behind the landscape is to condemn human mistreatment against nature. This explains why artists use caged vines, marking the limit between park and the city (Vercelloni and Vercelloni 2010: 255). In this same design, trees have been planted disorderly forming a small forest, reminding us of the natural organization of the forest ecosystem by comparison and bringing it close to the chaos in the cities, thus, building a wild urban forest (Vercelloni and Vercelloni 2010: 252).

The twenty-first century garden is leading us to rediscover ecology. In mid-twentieth century with the proposal of natural spaces in the city, the concept of Urban Ecology is born. With this new view, we discover that the city has many niches and ecological spaces; some of them are the garden and the green areas of the city. Each space in the city has been, in a way, conquered by nature. However, not only the vision of the city has changed but also the vision of agricultural spaces, revaluing them as cultural spaces and incorporating them to garden design. In part of Europe one of the problems is that the forest has reconquered the agricultural space; this has been used to value and facilitate the reconquest of plants and city ecosystems as well as to integrate agricultural systems to the design of parks.

Design in the twentieth and twenty-first century intents to recover the natural design of landscape looking for its creative strengths, in this case we will refer to the river basin as a force submitted by the city and dominated by its architecture and residences. This idea should be considered in order to incorporate the Rimac river basin to the design of Lima city. The design of "Back to the future" park (2004) in Seoul, intends to rediscover the history of space, and concretely of the basin. It could be useful to consider this design in order to integrate Rimac river. This design restores the origin of Cheonggyecheon river, by removing the cement that covered the river's natural source and makes it an essential part of the design, forming an ecological corridor of around 6 kms long (Francis and Chadwick 2013). A group of low-income inhabitants had settled along the river mouth, so the channel was covered by a highway to resolve the problem of pollution. However, in 2004 this space was uncovered and designed. This design intends to be a symbol of collective memory and make a historical rediscovery. Today along these 6 kms, the design reveals the history of the area; It rediscovers and gives value again to places where people would wash their clothes forming "natural pools". This space is currently one of the most beautiful in Seoul and is often visited. On the other hand, with the same idea of landscaping, we can see the reconstruction of an old factory in Bochum city, located on river Ruhr basin, Germany. It was highly polluted in the twentieth century but now we can find a cultural centre united to the park. In this restoration, nature reconquers space. Both cases should be considered to make a design that includes the Rimac river basin but considering its own historical background.

This search for nature includes also the history of the use of agriculture as a safe and peaceful natural space, like wheat fields. Since the twentieth century Frank Lloyd Wright (1867–1959) tries to go beyond the separation of the aesthetic garden seen from architecture, incorporating the agrarian landscape to the design (Vercelloni and Vercelloni 2010: 240). In Europe this type of design becomes important. It is about rediscovering the size of cultivated fields. This design reveals large homogeneous expansions where dimension enable the recovery of freedom of space in opposition to the small spaces of the polluted and overpopulated cities. This concept was followed in park Tempelhoff, a former airport in Berlin, Germany now turned into a park where people meet; this park recovers space by not even removing the runways. Currently, it is used as a space for entertainment. It is a return to large dimensions of open space and at the same time a reconquest of natural space. It is a minimalistic concept of design that recovers the conception of lost space. This can also by observed in the design of Potsdamerplatz in Berlin, Germany (see Fig. 4.12).

Even though the incorporation of agrarian landscape is still not an issue here, it is interesting for Peru in the twenty-first century that our agrarian tradition and space in the highlands gives us a very different vision. However, it also vindicates the agrarian landscape in the context of rural abandonment and loss. In current conditions, the Andean platforms are often abandoned because of the work that their maintenance demands (Yakabi 2018). In this sense, the Andean platforms represent an invaluable and unique landscape and cultural richness. For this, the integration and look of the migrant population gives us a new vision. In this sense, the revaluation of traditional agricultural spaces represents a challenge for landscaping in Peru as part of cultural revaluation and identification with space of the population living in Lima today.

Peru's difficult geology and orography gives a great potential for Peruvian landscape design. In this sense it is interesting to remember the concepts given by Patrick Geddes (Sects. 1.4 and 2.1), who tries to perceive and rediscover the geological aspects of landscape and combine them with the history and culture of the space, looking to connect the threetime dimensions: geological, ecological and historical. It could be interesting to express the research on our view of the desertic landscape in the great dimension of the landscape as well as the complex geography of the Andes. Here, design can be intertwined with the contrast of geometrical and organic forms of the landscape, what is natural and what is built. It is important to highlight the design of André Citroën park (Sect. 1.3) where a cypress wall opens a space to a freer design with disorderly nature. Likewise, in Seoul square, located opposite to Seoul National University, there is a contrast between trees and pavements, with curvy borders that form benches which invite people to sit and participate. This park is visited by many young artists and university students. It reminds us of the role of Spanish squares as places to meet and socialize as well as the complexity of spaces with Andean platforms and mountains opposed to the desert.

Even though it is still technically difficult to use gardens on green rooftops in Lima which has flat roofs without drains, green roofs give a new dimension to space from where we can contemplate

Fig. 4.12 Potsdamerplatz, Berlin, Germany. *Author* Ana Sabogal

the city like spectators with the vision of a flying bird, dominating space and also participating as spectator in the city that reminds us of the vision of spectator from the mountain in the highlands which could be expressed in an interesting design.

Retaking the subject of gardens in the ancient Lima and in general in Peru, there are many slummed, overexploited and deteriorated spaces that should be recovered. This could be done with small indoor orchards which depending on the space could have small fruit trees or even medicinal plants with a long tradition in Peru. A great creative effort as well as agronomic technics are required to recover these areas. The time dimension is a very important element in design. It is achieved when nature gains its own space and herbs start growing spontaneously, gaining a spot in the design, breaking the cement and reconquering the city. This idea could be used to rescue the lost orchards in Lima city.

On the other hand, another subject that can be considered in connection with rural migration is communal gardening in parks, planting not vegetables but medicinal plants since vegetables are difficult to grow in high temperatures and can be attacked by pests. These spaces could be a key fragment in the design of the space. Gardening could integrate the activities of the park, taking the idea of communal gardening in Europe. It could be managed by the municipalities with participation of citizens.

It is essential to integrate the vision of ecological corridors into the design so the city can be connected and integrated both socially and ecologically, taking the idea of the ecological corridors, specifically the one along the Spree river in Berlin (see Fig. 4.13). We can find this design in a centrical area of Berlin, and along it there are games for children, straight paths and a bike lane that gives us the sensation of space with the view of the city, from the riverbank as a spectator. Corridors allow us to follow the lines of space. In the case of Lima, it would be along the ocean front like the Costa Verde corridor, from where we can have a view of the ocean for kilometres in a continuous line, giving us a long vision of space. This corridor also connects land and ocean, giving us the freedom of an immense and empty space.

4.5 Species Used in Peruvian Gardens Throughout History

Lima city has very few native species; let's remember that most of its area is in a desertic ecosystem. Therefore, most are introduced

Fig. 4.13 Ecological corridor along the Spree river, Berlin, Germany. *Author* Ana Sabogal

species. So, when studying parks and gardens it is difficult to think about native species. The plants used in parks come from municipal plant nurseries that give priority to species of easy reproduction and fast growth.

Depending on the times, we can identify different arboreal species that predominate in different parts of the city. Among the plants commonly found in colonial times we have the geranium (*Pelargonium zonales*), fragrant plants like jasmine (*Jasminum officinale*) and trees like magnolias (*Magnolia grandiflora*) as well as fruit trees like fig (*Ficus carica*).

In Lima in this century we can find the yellow trumpet bush or huaranguay (*Tecoma stans*), which is about to become typical of the last decade, the Californian fan palm (*Washingtonia filifera*) and the bottle palm (*Hyophorbe lagenicaulis*) popular in the first decade of the twenty-first century.

In the twentieth century, the typical tree in the last decade (90s) was the *Ficus benjamina*, whereas the eucalyptus (*Eucalyptus camadulensis* o *Eucaliptus globulus*) was common in the 80s. The 70 s was marked by fruit trees like the mulberry (*Morus alba*) and in the 60s we could find the casuarina tree (*Casuarina equisetifolia*). The tipa tree (*Tipuana tipu*) was common in the 50s whereas in the 40 s the ficus (*Ficus elastica*) spread as ornamental plant. Table 4.1 lists the species used in Peruvian gardens throughout history. It is interesting to note that most species are introduced species.

Table 4.1 Species used in peruvian gardens throughout history

Species used in peruvian gardens throughout history		
Species	Period	Origin of species
Geranium (*Pelargonium zonales*)	Colonial times	Sud Africa
Jasmine (*Jasminum officinale*)		Central Asia
Magnolias (*Magnolia grandiflora*)		USA: Texas-Florida
Fig (*Ficus carica*)		Sud Asia
Ficus (*Ficus elastica*)	1940–1950	India
Tipa tree (*Tipuana tipu*)	1950–1960	Sud America
Casuarina tree (*Casuarina equisetifolia*)	1960–1970	Australia
Mulberry (*Morus alba*)	1970–1980	Central Asia-China
Eucalyptus (*Eucalyptus camadulensis*)	1980–1990	Australia
Eucalyptus (*Eucaliptus globulus*)		
Ficus benjamina (*Ficus benjamina*)	1990–2000	Sud Asia
Fan palm (*Washingtonia filifera*)	2000–2010	USA: California
Bottle palm (*Hyophorbe lagenicaulis*)		Mauricio island
Huaranguay (*Tecoma stans*)	2010–2020	Sud America

When walking around Lima, it is interesting to notice that you can determine when the park was established by the arboreal trees in it. As we can see, the choice of species is normally based on fashion since most species are not native but introduced for ornamental reasons. It is difficult to change habits related to planting species. Characteristics of species and fashion with some differences based on climate can be seen in other cities. In the last decades we can observe more awareness when selecting native species. The problem is that many native species, especially those from the highlands, are difficult to breed, yet, much effort has not been put forth to do so. Whereas plants from dry areas have deep roots so they are not suitable for the city, the ones from the highlands do not adapt easily to warmer ecosystems. Finally, the plants from the jungle are not common and most of them need frequent watering. Therefore, species should not be selected for breeding mainly because they are native, since urban ecosystems are anthropic, but for their agronomic characteristics.

City tree roots should be moderate in size. They should not be deep nor superficial. It is important to remember as we already mentioned in Chap. 3 that superficial roots can break sidewalks, whereas deep roots can break the pipes. Species selection should consider adequate formation pruning (see Chap. 3).

References

El Comercio (2019) Quinta Heeren un recorrido por la casona más misteriosa de Barrios Altos https://elcomercio.pe/vamos/consejos-de-viajes/quinta-heeren-un-recorrido-por-la-casona-mas-misteriosa-de-barrios-altos-fotos-noticia/. Revised: 3.01.2020

Endlicher W, Kraas F, Krajewski Ch (2012) Einführung in die Stadökologie. UTB, Stuttgart, p 272

FAO (2014) Organización de las Naciones Unidas para la Alimentación y la Agricultura. Ciudades más verdes en América Latina y El Caribe: Un informe de la FAO sobre agricultura urbana y periurbana en la región. Rome, p 92. https://www.fao.org/ag/agp/greenercities/pdf/GGCLAC/Ciudades-mas-verdes-America-Latina-Caribe.pdf. Revised: 3.01.2020

Francis R, Chadwick M (2013) Urban ecosystems: Understanding the Human Environment. Routledge, USA, p 220

Guías Expreso (1998) Paseos por la ciudad y su historia. “Inmigrantes Extranjeros”. La Quinta Heeren. Banco Sudamericano. Lima. Guías expreso (18): 241–243. https://www.caretas.com.pe/1998/1528/quinta/quinta.htm. Revised 17.03.2004

INEI (2015) Instituto Nacional de Estadística e Informática. Perú Anuario de Estadísticas Ambientales 2015. Lima, p 594

INEI (2013) Instituto Nacional de Estadística e Informática. Provincia de Lima compendio Estadístico 2011–2012, Lima, p 507

INEI (2008) Instituto Nacional de Estadística e Informática. Censos Nacionales 2007: XI de población y VI de vivienda: Perfil Sociodemográfico del Perú. Instituto Nacional de Estadística e Informática, Fondo de población de las Naciones Unidas, Programa de las Naciones Unidas para el Desarrollo (PNUD), 2nd edn. Lima, p 474

Lausent-Herrera I (1991) Pasado y presente de la comunidad japonesa en el Perú. IFEA, Lima, p 79

Liebenthal A, Salvenini D (2011) Promoting Environmental Sustainability in Peru: a review of the World Bank Group's Experiences (2003–2009). World Bank, Independent Evaluations Group. IEG working Paper 2011, N°1. Washington DC

Lima Cómo Vamos (2014) Evaluando la gestión de Lima: quinto informe de resultados sobre la calidad de vida https://www.limacomovamos.org/cm/wpcontent/uploads/2015/01/EncuestaLimaComoVamos2014.pdf. Revised: 17.04.16

Municipalidad de Lima, OEI (Organización Estados Iberoamericanos para la ciencia y la cultura) (1998) Barrios altos tradiciones orales

Pollan M (2010) The food movement, rising from The New York Review of Books. In: Wheeler S, Beatley T (eds) (2014) The sustainable urban development, 3rd edn. Routledge, London and New York, pp 339–343

PUCP (2016) Pontificia Universidad Católica del Perú. Instituto Riva agüero. 2016. Conferencia “La Quinta Herren y su jardín japonés. Historia de Tatsugoro Matsumoto migrante japonés al Perú y México”. https://ira.pucp.edu.pe/actividades/conferencia-la-quinta-heeren-y-su-jardin-japones-historia-de-tatsugoro-matsumoto-migrante-japones-al-peru-y-mexico/ Revised: 23.11.2019

Real Academia Española (2019) Diccionario de la Real academia Española. https://dle.rae.es/?id=MMXffef. Revised: 3.01.2020

Sabogal A, Cuentas MA, Tavera T, Varga F (2019) Espacios Públicos: Estudio del distrito de Santiago de Surco en Lima. Perú. Revista Kawsaypacha 3 (2019):105–138

Vercelloni M, Vercelloni V (2010) Geschichte der Gartenkultur von der Antike bis heute. WBG, Darmstad, p 275

Vernal F (2000) Ese lugar dentro del Barrio. "Touring", La revista del touring y automóvil club del Perú. 4 (13): 13–15

Yakabi k (2018) El abandono de los andenes de la comunidad campesina de San Juan de Iris. Escuela de Posgrado Pontificia Universidad Católica del Perú, Huarochirí

Zabalbeascoa A (2017) Hacia un jardín del siglo XXI. El País 26.12.2017, p 29

5 Park Typology and Legislation

Abstract

The purpose of this section is to review and discuss current park typology based on Peruvian regulations, followed by a proposal to change the existing typology. In this sense, we will analyse existing parks and the way they work. We will see that even though current park typology is complex, it does not pick up the complexity of Peruvian reality and is focused on Lima city. One of its weaknesses is the dispersion of functions. This section reviews existing Peruvian legislation covering public spaces and parks, as well as providing relevant examples and proposing future changes. It analyses the current typology for parks, and it proposes a new one considering the park's characteristics and patterns. This chapter contributes to the Development Sustainable Goal 11 and 16. It is oriented to encourage accountable and inclusive institutions, that promote a more inclusive and resilient city.

Keywords

Park typology · Legislation

The purpose of this section is to review and discuss current park typology based on Peruvian regulations, followed by a proposal to change the existing typology. In this sense, we will analyse existing parks and the way they work. We will see that even though current park typology is complex, it does not pick up the complexity of Peruvian reality and is focused on Lima city. One of its weaknesses is the dispersion of functions. This section reviews existing Peruvian legislation covering public spaces and parks, as well as providing relevant examples and proposing future changes. It analyses the current typology for parks, and it proposes a new one considering the park's characteristics and patterns. This chapter contributes to the Development Sustainable Goal 11 and 16. It is oriented to encourage accountable and inclusive institutions, that promote a more inclusive and resilient city.

5.1 Park Typology and Regulation of Parks in Lima

It is difficult to distinguish parks from public spaces. The difference is accentuated with the design of public spaces by the space planners in the twentieth century. Green areas are part of public spaces, being the plants their main elements of design, However, not all public spaces have green areas.

In Peru the legal difference between public spaces and green areas is not clear. In 2006, Standard G.040 of the National Building Regulations set out the first explicit definition of a public space in Peru, as a surface for public use

A. Sabogal, *Urban Ecology*, Sustainable Development Goals Series,
https://doi.org/10.1007/978-3-030-69905-5_5

made up of vehicle and pedestrian routes, parks, and plazas (Ludeña 2013: 43). In, 2011 the National Building Regulations incorporated the urban dimension as part of the concept of public spaces, focusing on walkways and roads as well as parks and plazas (Ludeña 2013: 34). Pursuant to existent laws passed by the "*Ministerio de Vivienda, Construcción y Saneamiento*" and the "*Servicio Nacional de Capacitación para la Industria y la Construcción*" (2006), a public space is considered "a surface of public use with the purpose of entertainment or circulation" (Sabogal et al. 2019). When looking into laws about green areas we find disperse information and functions, the "*Ministerio de Vivienda, Construcción y Sanamiento*", the "*Ministerio del Ambiente*", Lima Park Services, Lima municipality and municipalities in general are responsible for regulating on the subject. Regulations in force in Lima should be distinguished from other regulations in the country where park implementation and distribution of functions is competence of their own municipalities; however, few or no laws have been passed except for Trujillo city. The "*Reglamento Nacional de Edificaciones*" (National Building regulation), in force in all Peru, defines a park as an open space of public use for passive or active entertainment, with predominantly green areas.

The history of green areas in Peru dates back to the twentieth century and centers on Lima; they are conceived of primarily as open areas that combine greenery with public access. In his review of the legislation, Ludeña (2013) notes that the first official document to consider public areas dates back to 1949; Formulated by the National Office of Planning and Urbanism (ONPU), it defines three types of open areas: rural green areas, made up of cropland; semipublic green areas, made up of private clubs; and public green spaces, made up of parks. It also incorporates an Extraurban Recreational System comprised of beaches and valleys no more than 45 min from Lima (ONPU 1949 cit. Ludeña 2013: 40). Thus, the notion of "open area" is long-standing but still highly relevant to the design and planning of the city. In 1958 the ONPU defined open and recreation areas for Lima, based on four categories: City; sector; neighborhood; and extra-urban recreation, including beaches and valleys (Ludeña 2013: 39). In 1964, the Board of National and Zonal Parks was founded; this was followed by the Metropolitan Lima–Callao Development Plan, which introduced two categories of extra-urban areas: urban, consisting of parks; and regional, consisting of beaches and valleys (Ludeña 2013: 40). In 1977, the Metropolitan Lima General Zoning Regulation introduced specific regulations for the Costa Verde, marking an interesting evolution in the concept. In 1985, the Territorial Conditioning, Urban Development and Environment Regulations incorporated conservation areas and protection areas, the latter including rivers along with their banks and beds (Ludeña 2013: 42). In 1995, the National Construction Regulations incorporated certain components of the road system, such as expressways, avenues, boulevards, parkways, *jirones* (short, pedestrianized streets), streets, and alleys (Ludeña 2013: 43). in 1995, zonal parks were legally constituted in Peruvian legislation (SD 04–95 MTC) (Ludeña 2013: 43).

In 2003, Municipal Ordinance of Metropolitan Lima No. 525 of 2003 defined the following categories of green areas: metropolitan parks, zonal parks, local parks in urban sectors (including sports fields and parks, neighborhood parks, and neighborhood sports areas), local residential group parks, natural historical monuments, plazas, *plazuelas* (small plazas), promenades, boulevards, embankments, natural forests, planted forests, central gardens, and lateral gardens (including public roads and intersections) (Ludeña 2013: 43). This was the first time that the category of natural historical monument was added, and that a distinction was drawn between natural and planted forests (Ludeña 2013: 43).

Another important step was the addition of organic gardens (*biohuertos*) to the regulations, and their promotion. Ordinance 1629 of 2012, for the Promotion of Urban Agriculture (El Peruano, 24 Sep. 2012) promotes urban agriculture in public and private spaces and the conservation and utilization of agricultural areas in the Chillón basin.

Pursuant to Municipal Ordinance MML 1852 de 2014, Green Areas are spaces where plant species can be placed to be managed by local or metropolitan municipalities (El Peruano 2018). Based on this regulation, green areas are public. Management of public places differs depending on size, importance and municipal space. Thus, green spaces can be managed by Lima Municipality who based on the size and importance could delegate management to Park Services in Lima (SERPAR), when the green area expands across several districts or to a district municipality when it is located only in only one district. Then, fore green areas there can be municipal plans, district plans and a guideline for management of green areas. Furthermore, ordinance 1852 includes detailed regulations for pruning.

Municipal resources for green areas depend on public priority that is determined together with the population of the district. On the other hand, SERPAR reports to Lima Municipality who provides it with a budget. SERPAR is responsible for nine metropolitan parks; one of them is the "Parque de la Exposición" as well as zonal clubs, sport and cultural schools, zoos and botanical gardens in Lima (SERPAR 2019).

In Peru, great part of the cities on the coast are located near the waterfront which is also a public space. The General Municipal Law 27,972, published on May 27, 2003, determines that Public Areas are of public use and domain (Congreso de la República 2017). Likewise, the areas on the waterfront are regulated by the law 26,664 passed on September 8, 1996 which determines a two-hundred-meter strip for public use after 50 m of beach as long as the beach has geographic continuity. Beaches on the coastline are also goods of public use and should be dedicated for public use (Congreso de la República 2017). Therefore, they are Public Areas. Finally, the same General Municipal Law 27,972 determines that municipalities have the faculty to transfer use of goods for exploitation so that these goods can serve social interests and needs, giving a step backwards and neglecting their public nature.

The ordinance for the Preservation and Management of Green Areas published on December 28, 2014, MML 1852 establishes the current typology for Green Areas of public use, determining the following categories:

- Urban and Peri-urban parks comprising different type of parks such as Metropolitan, Zonal, Local, Riverside, Cultural, Hilly, Landscape Protectionist, Agrarian, Botanical, Lineal, Zoo and Natural Forests parks.
- Complementary Green Areas including urban orchards, green roofs and others.
- Areas of Environmental Reserves including natural areas (El Peruano 2018).

Urban and peri-urban parks, depending on the size and ownership of the green areas can be categorized in three groups:

- Metropolitan parks: They are usually very large, and their purpose is landscape value and environmental benefits. Since these parks have a specific function, they are going to serve the metropolitan population all together. Public activities such as parades, dances, etc. are held in these parks.
- Zonal parks: These are large green areas located in residential zones but in more than one district. They can have equipment and provide environmental services. They are very large with multiple functions. Its purpose is entertainment of the population in suburban areas in Lima city. It provides services of entertainment such as theatre, walks, sporting fields among others.
- Local parks: Their dimension is based on the size of the district and it is managed by the local municipality (El Peruano 2018).

Since urban and peri-urban parks consider nature and ecology, they include Riverside, Cultural, Hilly, Landscape Protectionist, Agrarian, Botanical, Lineal, Zoo and Natural Forests parks. Linear parks are entertainment and conservation corridors whereas Agrarian parks are areas of agrarian production inside or outside the city. It is important to be aware that Urban Orchards are included in the category of Complementary Green Areas different from the agrarian parks that only have an agrarian purpose (El Peruano 2018). Finally, the zonal parks in Lima City are large parks managed by SERPAR.

Ordnance 1852, for the Conservation and Management of Green Areas in the Province of Lima, established the following management plans for green areas within the municipality:

1. Metropolitan Plan for Green Areas, including the assessment of green areas and development plans over a period of ten years.
2. District Plans for Green Areas, drawn up by municipalities every five years.
3. Metropolitan Inventory of Green Areas, overseen by the Municipality of Lima and renewed every three years.
4. District Inventory of Green Areas, renewed every three years.
5. Technical manuals and guides for the protection, management, design, and conservation of green areas and urban trees, prepared by the Environmental Department of the Municipality of Lima. Compliance is mandatory throughout Lima.

The rules set out in these manuals include interesting aspects related to the environment, such as: selection of flora species that require little water; substitution of grass species with those that consume less water; recommendations not to use species susceptible to pests or that require high maintenance; regulation of pruning, with instructions for adequate completion; requirements that forest nurseries be established in Lima's municipalities in order to stock green areas, with priority given to native species; and the authorization of public—private partnerships for the implementation of public areas. The regulations include a series of sanctions aimed at controlling indiscriminate pruning of trees, damaging trees, among others. Although monitoring trees and pruning is fundamental to good park management, it is not mandatory. Lima does not currently conduct a tree census, though the following districts do: Central Lima, Miraflores, San Borja, San Isidro (Lima Cómo Vamos 2015), all of which have relatively high purchasing power.

This same ordinance establishes that a maximum of thirty percent of each district's park surface area be for active usage, and that no less than seventy percent be under green coverage; meanwhile, zonal parks were to designate eighty-five percent of their area as free space, and seventy percent as green areas.

An important management instrument is the Lima and Callao Metropolitan Area Plan, 2035 (PLAM 2035). The PLAM 2035 proposes various advances in the environmental management of Metropolitan Lima, including: sustainable urban transportation, prioritizing systems for pedestrians and bicycles (ONU Habitat 2015: 23); creation of hill conservation areas and environmental corridors; and recovery of the ecological structure of the Chillón, Rímac, and Lurín rivers (ONU Habitat 2015: 23). This involves the proposed recovery of hills, wetlands, coastline, and valleys to overcome the lack of green areas in Lima and Callao by 2035 (ONU Habitat 2015: 40).

5.2 Proposed Typology for Parks

Lima as one of the first cities of the viceroyalty as well as Trujillo and Piura have a historical tradition of park and garden development, even though the first gardens were indoors in convents and mansions. Squares are important ceremonial centres since these cities were founded. It is only at the beginning of the twentieth century with the development of parks or specifically with the "Parque de la Exposición" that parks are open but later in the 60s they are closed again by fences because of socio-cultural problems in the city. Considering history and the current development of the city, as well as the typology previously described and field studies a new classification for parks is proposed. This includes the following categories:

- Zonal park,
- Monumental park,
- Historical park,
- Nature conservation park,
- Modern park,

- Neighbourhood parks,
- Ecological Corridor.

As follows we describe, support and give examples of each categories.

5.2.1 Zonal Parks

The current definition given to zonal parks already described in the above category is the one that will be used in this proposal since it is important to have large open and ample spaces that serve the citizens for entertainment and sports. A Zonal Park represents a proposal of ample activities for inhabitants of all ages and from different districts. This category has an important role of development and integration of the city.

The zonal parks were created in the 70s when Lima city was in expansion due to migration caused by the change in land ownership pursuant to the Agrarian Reform law. The Zonal Parks pick up the idea of the great parks from the beginning of the twentieth century: They are parks of great dimension, and their purpose is entertainment of the population in suburban areas of Lima city. These areas can include sports facilities, swimming pools, strolling areas, mini zoos among others such is the case of the Zonal Park of Chimbote. Most of the visitors to these parks are low-income families which makes them essential spaces for entertainment and sports. Even though these Zonal Parks are frequently used by the population, furniture has not been renewed and they have very little resources which are poorly managed; this affects the functionality of the park and puts its own existence at risk.

We will describe the Zonal Park of Chimbote as an example. It has 34,5 hectares and it is subdivided in multiple spaces Therefore, there is a great movement in the park, which is used by a diversity of people, mainly by families with children and to a lesser extent, by young people. In spite of this, some areas of the park are underused and neglected. Many spaces are sublet due to bad management of the park as well as lack of resources and reinvestment. This is a general problem both for Zonal Parks as well as parks of other categories. The main reason is that there is no awareness of the magnitude that green areas require a maintenance budget. Most municipalities invest in the inauguration of the park. However, they do not keep an adequate budget for its constant renovation and maintenance. In the Zonal Parks we can find a lot of street vendors or small restaurants that could be a good source of money, but currently they contribute very little.

Among the spaces in the park, we can find an area for children's mini golf that does not work as well as a deteriorated acoustic shell with moth-eaten wood and old, broken and dirty plastic; it is currently used only to pass radio music. Furthermore, there is an area for children's games which are in bad conditions (see Fig. 5.1) and a boat museum with very old fish kept in formalin in need of renewal. There is also an olympic-sized pool. Among the most frequently used zones we can find an area with tables for snacks as well as a leased area with bumper cars (*Chachicar*). In the latter, the music is too loud, and the eucalyptus are not well kept so they are a danger to pedestrians. This area is the most visited and enjoyed. Another circuit that unifies the park is the train which is in good conditions and goes through natural spaces connecting them with the most frequently used areas.

As we can see the park was designed with creativity and adequate space demonstrated by the design and implementation of the Boat Museum, very significant for a fishing city like Chimbote; it is an educational and emblematic space (see Fig. 5.2). There has obviously been a high initial investment in educational material, but it has not been given maintenance. If we examine the swimming pool, we arrive to the same conclusion; lack of maintenance and reinvestment; the olympic-size pool has a crack in the middle that does not allow its use. Furthermore, it is too deep and dangerous (3.5 m).

The design for use of water has not considered water flow comprehensively. The park has a lot of natural water, including a channel surrounding it that comes from Santa river. However, this

Fig. 5.1 Chimbote Zonal Park, playground. *Author* Ana Sabogal

Fig. 5.2 Chimbote Zonal Park, Boat Museum. *Autor* Ana Sabogal

water flows into the ocean without being used. The water for the park comes from wells, two of which cannot be used. Their motors are with sand and the pumps are 60 years old; the sub-station has not been given maintenance for 20 years. There are natural lagoons filled with debris. One of the lagoons has been backfilled and covered with stones but has not been given maintenance so the edges are filling with aquatic plants. The lagoon inside the park has collapsed. It was previously used as sewage for the bathrooms but currently there are new bathrooms with a septic tank.

Like most Zonal Parks, it has a lot of animals and an area for a zoo in spite of not having an official permit as kennel zoo. Many animals are scattered around different parts of the park. Among the animals, we can find asses forming a big group in very poor conditions, infested and without care from a veterinarian. However, the asses are offered to visitors for children's rides. In the area of the lagoon we can find turtles and parrots and in the zoo the animals are in overcrowded cages. Among them we can find toucans, Humboldt penguins, ducks, two woolly monkeys, two ocelots, spider monkeys, iguanas and opossums. These animals are not given enough care by veterinarians who only attend every six months. We also find animals in the artificial lake used for boat rides which include ducks, penguins, carps and lizas.

The design of the park has not included the park's natural spaces which comprise a natural area of water springs, abandoned by park management, which is surrounded by neglected old trees that have been invaded by Chinese grass, cattail and high groundwater. This area is used by outsiders who lease it in order to get cattail for mat weaving.

Regarding management of green areas in the park, we observed no forestry management, which poses a great hazard since trees in the area are eucalyptus and they can fall, or their branches can break at and time. In general, the trees in the park are 40 years old and they have not been renewed or given proper maintenance; good management is required to keep proper control of the trees in order to replace them in a timely manner. Tree pruning and maintenance is necessary to ensure safety for pedestrians. Currently, trees are facing disease so they are particularly risky since branches can fall over pedestrians at any time. Furthermore, the space has been invaded by *Ipomoea alba* which are in competition with the trees, especially along the railway for the train. Just like in any park, an inventory of trees should be kept with a file for each one; the park requires time and a budget for maintenance.

When studying the problem, we can see that one of the recurring difficulties is park management, which is a similar situation in all Zonal Parks. In general, there is inadequate planning, reinvestment, maintenance as well as lack of personnel. In this particular case, the park has 14 people working full time, which is not enough to give maintenance to the 35,4 hectares of park. The plant nursery, which is counted upon for plant reproduction and planting has only two workers; clearly there are not enough people. This situation leads to the lease of farming areas in the park and the sale of grass. However, it does not allow to designate space in the park for grass maintenance and replanting, thus, these activities become more expensive. On the other hand, the restaurant which has been sublet is not given proper maintenance nor is there evidence of reinvestment and it lacks proper management of the space, which is the case for all leased spaces in the park. This situation makes spaces deteriorate even further and will have to be repaired at the State's expense. We see then that there is an intention to get more income by leasing spaces, but this income turns out to be meagre, especially considering that there is no investment in the spaces leased. The park has severe safety issues since the fence surrounding it is partially torn down which for safety reasons has been completed with debris. Finally, the park has not escaped social problems and houses a great number of children that have been abandoned by their families. We can see a public space with a great network of social problems which escape the grasp of park management; reinvestment is inexistent nor is there a long-term projection and least of all planning.

One of the problems with the design is that it does not consider natural spaces. However, the great dimension of the park as well as its natural ecosystems entitle it to a landscape design which takes into account these natural spaces. The natural lagoon in the back could be cleaned and used for pedal or fishing boats. The lagoons should consider the use of filtrating species for the ecological maintenance of the space. The species of flowers used could be the achira (*Cana edulis*), the flower cartridge or the calla lily (*Zantedeschia aethiopica*). It would also be interesting to have an area for bird watching where the cattail is, in order to promote bird nesting and create a circuit for bird watching. To make the space around the artificial lagoon natural again, native riverside mountain trees could be planted around it such as willows (*Salix humboltiana*), elder trees (*Sambucus peruviana*) or baccharis (*Baccharis sp.*). Moreover, Peruvian ducks could be introduced to help build a trophic chain since they would feed on existing fish (see

Fig. 5.3 Natural Outcrops with abundance of cattail, Chimbote Zonal Park. *Author* Ana Sabogal

Fig. 5.3). We also propose to leave a natural hilly area and develop dunes which could have native plants from the coastal dunes.

To improve landscaping, considering the dimension of the park we also propose to develop "alamedas" which would make the landscape uniform. The "alamedas" could be formed by tipa trees and cedars (*Cedrela odorata*) since they are tall trees with a wide crown. Besides, they are native and already present in the park. Finally, along the train railway there could be a circuit for environmental education which could include the natural ecosystems.

The problems described are similar in other Zonal Parks both in Lima and in other Peruvian cities.

5.2.2 Monumental Park

The typology of Monumental Park corresponds to extensive parks intended mainly to represent the city. This typology is linked to the history of the city. That is why it is given the name of "monumental", making a reference to their importance as parks of great dimension with a historical and institutional scope. The "Parque de la Exposición" recently named "Gran Parque de Lima", mentioned before is a clear example of this typology.

It is important to point out that the "Parque de la Exposición" was built in the outskirts of Lima city and after time some time it became part of the centre of the city, dividing in several smaller parks. Let's remember that the park was created to replace the city wall, changing the city's personality and giving it a new look, more modern, that picks up the idea of world culture.

5.2.3 Historical Park

The typology of Historical Park corresponds to those parks that were founded together with the city, as part of its planning or in the process of development. In the specific case of Lima, these parks were created during the viceroyalty period and have gone through great changes and transformations as a consequence of city growth. However, they are still important for their use, for being part of the city's composition and for combining green areas and art. Most of these parks have very valuable sculptures; statues or fountains are never missing. Many of them are private spaces which should be revalued and become public.

Lima city has different historical parks. One of the parks is the "Paseo de los descalzos" previously described. The park has beautiful sculptures along an "alameda" typically used by the viceroy for his walks. The role of French style parks served as an element of power so that the high classes could take walks as well as assist to musical and public events. This category includes important private historical monuments such as the Quinta Herren also described above which have great historical value and therefore its role is that of the typology of historical park as long as these spaces are actually turned into parks. This is also the case of the different orchards in convents such as the previously mentioned "Leprosorio de San Lázaro".

5.2.4 Nature Conservation Parks

Park typology for nature conservation includes spaces that have the function of nature conservation, research, environmental education and sustainable tourism. These parks are especially important in big cities like Lima since the city's size makes it difficult for inhabitants to know the natural ecosystems contained in it. On the other hand, they are also important as sanctuary for flora and fauna. This typology is significant in the context of ecological abundance and multiplicity of ecosystem corridors in Peru. Villa Marshes Nature Reserves corresponds to this category (see Sect. 7.3, Fig. 5.4). Although the Zonal Park of Chimbote previously described does not correspond entirely to this category (see Sect. 7.3), it could if spaces and ecosystems in the park are revalued as we already have proposed.

The only natural protected area in Lima is the Villa Marshes Nature Reserve. This space is in the south of Lima, located in Chorrillos district near The Panamericana Sur Highway. These wetlands are formed with water from Rimac and Lurin rivers and is connected to the other wetlands on the coast of Peru forming an important network of underground water along the coast. Wetlands in Villa are a sanctuary for birds on their migration route from the Andes and the north of America. This space is one of great biodiversity, especially of birds, it has an interpretation centre and it is easily accessed so it fulfils the important role of environmental education.

Another park that fulfils the role of environmental education is the "**Parque de Las Leyendas**", founded in 1964. This park has

Fig. 5.4 Natural Outcrops, Villa Marshes Nature Reserves. *Author* Ana Sabogal

prioritized the concept of environmental education and it includes historical and archeologic aspects as well as flora and fauna that represents each one of the natural regions in Peru. Finally, we can find a botanical garden in the "Universidad Nacional Mayor de San Marcos" which has practically been forgotten even though it is the first botanical garden in Latin America. However, it still has some important botanical species. This garden should be revalued to fulfil the role of providing environmental and historical education in relation to plant conservation and botanical classification.

5.2.5 Modern Park

This typology includes most of the new parks commonly found in Lima. Most of them are managed by district municipalities and others have been built by an independent initiative, some with external funding and others with government funding in coordination with municipalities or Lima Municipality in order to improve green areas in Peru and ensure a good quality of life for its inhabitants. The "Alameda de la Juventud", previously described, is in this category; it transformed an avenue in a green area improving the quality of life of inhabitants of a district that does not prioritize green areas due to the number of overwhelming problems in need of immediate solution such as solid waste management together with non-payment of municipal taxes, among others. In this category we can find parks based on simple sketches without much spatial design exclusively to meet municipal regulations regarding green areas. Such is the case of parks in Santiago de Surco district that have a simple geometric and homogeneous structure without much design.

A typical example of this typology is the "Parque La Alborada", located in Santiago de Surco district in the south of Lima (see Figs. 5.5 and 5.6). The park has a good location with proper road connections for both vehicles and pedestrians. For this reason, it has been used for many years to accommodate the so called "Bioferia" or market for organic food. Since then and thanks to this activity, it has become a point where district neighbours meet (Sabogal et al. 2019). Like several parks of the area, it has a children's playground, "alamedas", calisthenics area, and the distinctive Virgen in the middle of the park, common in all the parks of the district.

Fig. 5.5 Modern park, playground, La Alborada park, Santiago de Surco. *Author* Ana Sabogal

Fig. 5.6 Modern park, "alameda" area, La Alborada park, Santiago de Surco. *Author* Ana Sabogal

Currently, it also has a Citizen Safety Stand that makes neighbours feel secure and visit the park more frequently.

Another interesting park very similar to the one described before is the Marco Schenone Oliva park, also located in Santiago de Surco district (see Figs. 5.7 and 5.8). This park uses the typical scheme of the Modern Park typology with straight and diagonal lines intersecting it. In this park there are vaccination campaigns for dogs as well as traditional food fairs. Furthermore, it is frequently used by neighbours for hikes and sports or simply for dog walking. Groups of people get together for different activities, such as a capoeira group that practices the sport here as well as neighbours that get together to celebrate birthday parties for their children. Like the previously described park, it has a playground, the Virgen typical of the district and an educational area with Sen plants (*Cassia angustifolia*), which are used as medicinal plants to fight anemia. Both parks fulfil their role and are frequently used by neighbours.

Fig. 5.7 General view of Marco Schenone Oliva park; Santiago de Surco, Lima. *Author* Ana Sabogal

Fig. 5.8 Virgen in Marco Schenone Oliva park, Santiago de Surco, Lima. *Author* Ana Sabogal

5.2.6 Neighbourhood Parks

In most Peruvian cities in expansion, there are parks that can be considered in the typology of neighbourhood parks, which are the result of usual city dynamics. They are spaces planned for parks that are planted and care for by the people who live in front of the park. The municipality appears after the area is urbanized to plant some grass and maybe plant a tree. At that point, the park is cared for both by the municipality, that waters the park sporadically and by the population. In these parks many species are planted by the population such as fruit trees like avocado, bananas or medicine plants such as lemongrass (Sabogal and Martínez 2015). Even though these parks are planned on a municipal desk, their design is actually random due to the prior intervention of the population; the municipality respects the plants placed priorly by the neighbours. The species planted are usually disperse fruit trees and medicinal plants and only a few or no flowers. Once the municipality arrives, it plants and cares for the vegetation, but the population always cares for what they planted, making sure it is watered properly and harvesting produce to complement the population's diet. Usually the park has a geometrical structure and little diversity of species with very few ornamental trees that are usually the same ones around the city which is not the case for fruit trees. These parks compete directly with the municipality that allots the area and size based on the regulations for development of green areas in Lima. The parks are developed with the participation of the neighbours who frequently are responsible for the area in front of their house. Most Neighbourhood parks can be found in low-income areas in marginal urban zones and the produce obtained complements the local population's diet supplying them with minerals and vitamins. The park is also a space where neighbours gather to do activities like watering and harvesting.

Parks in middle-class districts frequently become zones whose access is restricted to inhabitants; railings, gates and other barriers are erected to prevent entry to non-residents, while "keep off the grass" signs are erected and cacti planted to restrict circulation (Ludeña 2013: 148). At present, these parks are highly used for temporary events such as organic product fairs, concerts, sports, dog contests, and others.

In most of cases, the neighbours choose the species they want and plant them initially, later the municipality plants more trees and grass (see Fig. 5.9). In many cases, these parks have religious statues and ornaments that are placed in the centre by the municipality. After this, the park is inaugurated officially with municipal speeches and it is given a name as a way to take power over the space. Neighbourhood parks have a simple architecture with a central sidewalk as well as secondary ones. Furthermore, the architectonic treatment uses straight, defined and rigid lines. Thus, spaces in front of the houses which are responsibility of municipal management, are developed with participation of the neighbours who many times are the ones that take charge entirely of the part in front of their house. They choose the species to be planted and then provide maintenance and harvest the produce (see Fig. 5.10).

Fig. 5.9 Neighbourhood park with banana tree in Cercado de Lima, city town of Lima. *Author* Ana Sabogal

5.2.7 Ecological Corridor

This typology is not frequently found in Lima. However, it is essential for the development of the city. Even though the name is not entirely appropriate, it is partially considered under the category of Urban and Peri-urban parks within the Linear parks defined as entertainment and conservation corridors (see Sect. 4.5.1). The ecological corridor is important for several reasons. On one hand, it enables inhabitants to take walks improving their health, and on the other hand, it fulfils the invaluable role of integrating social spaces. Finally, it enables the development of fauna and flora in the city uniting spaces as well as the population of flora and fauna, ensuring their vigour and survival in the city.

A very important example of this typology is the corridor along the Costa Verde described lines above. There are also smaller corridors that are also important since they fulfil the role of ecological corridor. An example of this is Arequipa avenue that unites several districts for

Fig. 5.10 Neighbourhood gardens in front of a house, Cerro San Cristobal, Lima. *Author* Ana Sabogal

Fig. 5.12 Ecological corridor along Pardo avenue, Miraflores, Lima. *Author* Ana Sabogal

several kilometres crossing the city from the centre to the ocean connecting with the Costa Verde corridor. On Sundays this corridor is used to do sporting activities such as cycling and walking. Another one is Larco Avenue located in Miraflores, that also connects with the Costa Verde corridor uniting spaces (see Fig. 5.12).

The development of the ecological corridor in Lima that goes along the coast and communicates the city with the ocean is another space with problems of conception and design. Lima municipality has invested a large budget to develop it but there is a contraposition of two incompatible functions: a boardwalk for pedestrians and sports

Fig. 5.13 Costanera ecological corridor along the sea, Lima. *Author* Ana Sabogal

next to a high-speed highway that makes access difficult and dangerous (see Fig. 5.13). However, it is an important area for the city as a natural space that properly developed can be a beautiful attraction in Lima city improving inhabitant's quality of life. Through this example we can see the complexity in the design of corridors which should consider and make ecological aspects converge with space management and connections to enable pedestrians to use them and assure that they fulfil their ecological and social functions.

References

Congreso de la República (2017) Proyecto de ley 1311/2016. Proyecto de ley Ley de gestión de espacios públicos. https://www.congreso.gob.pe/Docs/comisiones2017/Comision_de_Descentralizacioni/files/pl0131120170425.pdf Reviced: 01.12.2019

El Peruano (2018) Ordenanza que reglamenta la conservación y gestión de las áreas verdes de uso público en el distrito. Ordenanza 478, 25.04.18. Municipalidad de San Isidro. https://busquedas.elperuano.pe/normaslegales/ordenanza-que-reglamenta-la-conservacion-y-gestion-de-las-ar-ordenanza-no-478-msi-1642505-2/ Revised: 1.12.2019

Lima Cómo Vamos 2014 (2015) Evaluando la gestión de Lima: Quinto informe de resultados sobre la calidad de vida. https://www.limacomovamos.org/cm/wp-content/uploads/2015/10/ReporteAmbiente2014_virtual.pdf Revised: 17.08.20

Ludeña W (2013) Lima y Espacios Públicos, perfiles y estadística integrada 2010. PUCP, p 224

ONU Habitat (2015) PLAM 2035, sistematización del Plan del Área Metropolitana de Lima y Callao, 2035

SERPAR (2019) Servicio de Parques de Lima. https://www.serpar.gob.pe/quienes-somos/. Revised: 01.12.2019

Sabogal A, Martínez M (2015) A study of ecological corridors in two quarters of Lima: Chorrillos and Villa El Salvador. Perspect Global Dev Technol 14 (2015):587–596

Sabogal A, Cuentas MA, Tavera T, Varga F (2019) Espacios Públicos: Estudio del distrito de Santiago de Surco en Lima. Perú. Revista Kawsaypacha 3 (2019):105–138

Environmental Problems

6

Abstract

Cities typically have pollution problems which affect all its inhabitants. This chapter supports the need to consider concepts related to green areas as a necessity for the city. It describes pollution produced by a diversity of human activities in the city, differentiating between its effects on soil, air, water and climate. It distinguishes heavy metals and chemical products applied in parks and gardens which pollute the soil. It describes air pollution produced by motor vehicles as well as by the persistent use of organic matter. It highlights the function of trees as carbon dioxide absorbents, acknowledging that they cannot absorb all emissions which should be reduced. It describes problems related to water and ground water polluted with heavy metals and other pollutants, highlighting its effect on eutrophication and the destruction of wetlands. It emphasizes that temperature has risen in the cities. Furthermore, it describes the function of the ocean and wind in relation to the location of Lima. The description focuses on Lima city and on the search of solutions. This chapter aims to contribute to the enrichment of a city with better environmental conditions to improve the health of its inhabitants. It contributes directly to the Sustainable Development Goal 11 and indirectly to the goals 3.

Keywords

Environmental problems · Soil · Air · Water · Climate

Cities typically have pollution problems which affect all its inhabitants. This chapter supports the need to consider concepts related to green areas as a necessity for the city. It describes pollution produced by a diversity of human activities in the city, differentiating between its effects on soil, air, water and climate. It distinguishes heavy metals and chemical products applied in parks and gardens which pollute the soil. It describes air pollution produced by motor vehicles as well as by the persistent use of organic matter. It highlights the function of trees as carbon dioxide absorbents, acknowledging that they cannot absorb all emissions which should be reduced. It describes problems related to water and ground water polluted with heavy metals and other pollutants, highlighting its effect on eutrophication and the destruction of wetlands. It emphasizes that temperature has risen in the cities. Furthermore, it describes the function of the ocean and wind in relation to the location of Lima. The description focuses on Lima city and on the search of solutions. This chapter aims to contribute to the enrichment of a city with better environmental conditions to improve the health of its inhabitants. It contributes directly to the Sustainable Development Goal 11 and indirectly to the goals 3.

A. Sabogal, *Urban Ecology*, Sustainable Development Goals Series,
https://doi.org/10.1007/978-3-030-69905-5_6

6.1 Environmental Problems

With city growth, the number of environmental problems increases. Population density together with carbon dioxide emitted by factories during the industrial age and currently by vehicles have triggered the need for parks in the cities. Thus, at the beginning of the nineteenth century, English doctors determined the necessity for green spaces in the city. Subsequently, the World Health Organization regulated it. All this has spawned a change in the view of the park from a private space where the bourgeois went to take a stroll in the garden to a public requirement. So, gardens of the nobility have become parks of collective use by the citizens. Furthermore, there is also a change in the perception of the park from a poetic or aesthetic conception to a social one, in which parks should be useful and fulfil certain functions (Vercelloni and Vercelloni 2010). Natural limits can determine the urban development of green spaces which then blend with local history and development, determining the character of the city and its green areas (Whiston 1984: 64). The natural aspects that define and give character to Lima's development are the hills as well as water and ocean surrounding it.

One of the problems for green areas is the use of chemical, pesticides and fungicides. This problem is especially relevant in Lima city. The chemical substances have a diversity of characteristics. Their capacity of diffusion, chemical stability and capacity to transfer from one component to another depend on their composition. Once the chemical product reaches the environment a diversity of processes is triggered; firstly, it is transported in the environment; secondly, it is transferred from the product to other spaces. Then, for example, heavy metals like mercury, found in chemical products that are applied to city pests are transferred from the soil through percolation as it is carried by groundwater and finally it is transformed by integrating the trophic chain or it is carried away in the water to other ecosystems. The speed of transportation as well as the capacity of transformation also depends on the product's composition. This means, substances have a different degree of stability; the tendency to transform from one substance to another is called fugacity. Most chemical products, when transported or transferred keep their chemical properties, provided that the pH is not very different. Although many chemical transformations often occur in the air, biological transformations often occur in the water; finally, both types of transformation occur in the soil. There are multiple connections between water and air; the substances in water often transform from liquid to gas and end up in the air. A similar process occurs between the air and the water. Sun rays give energy used in processes called photolysis that enable the volatilisation of chemical products in the ground, soil or air (Fent 2003: 39). These processes occur in water treatment plants, stagnant water, or farming soil (Fent 2003: 39). If a substance begins a biological process the chemical product often transforms. Heavier chemical products that decompose with more difficulty end up as sediments which often become part of the trophic chain by effect of bacteria; in this way biological process is triggered.

We tend to divide nature in good or bad. However, not all changes in the landscape are negative; many of them contribute to improve urban life, lowering temperatures and cleaning the air, especially in a city like Lima that does not have natural green spaces. In Lima and in great part of Peru there are many crowded, overexploited, deteriorated and polluted green spaces that should be recovered in order to offer ecosystem services such as air cleaning. In Lima and in Peru in general, a large number of the green spaces are overcrowded, overexploited, deteriorated, polluted and in need of recovery; a considerable effort of imagination and agronomic technique will be required if they are to provide ecosystem services again. Lima's main problems in this regard include air pollution from particulate matter and greenhouse gases; soil pollution from cooking oil, chemical products such as detergents, pesticides, and petroleum; and water pollution from heavy metals.

Urban ecosystems, conformed by city parks and gardens, offer fundamental environmental services by storing carbon through plants, regulating soil erosion, controlling water floods and finally, regulating climate through vegetation as a whole by buffering the increase of temperature in the city. Green areas provide supporting services by enabling water and nutrient cycles. Furthermore, they offer cultural service by connecting societies and enabling the care and respect of the environment. Thus, to recover Lima, its air, soil, and water must be in good condition.

6.2 Soil

Soil in the city is very different from soil in the countryside. It is compacted by the buildings; pH is increased by the cement in buildings and foundations. Furthermore, the depth of groundwater is affected by the buildings. The weight of constructions disturbs the flow of air and water in the soil. Roads and sidewalks modify water and groundwater circulation greatly. City wetlands are the first to suffer the consequences of compaction (Francis and Chadwick 2013: 31). Asphalt heats strongly and produces evaporation. Green spaces and dirt lanes are particularly interesting since they enable life in the ground and helps regulate the climate in the city (Endlicher et al. 2012: 113). There are many environmental technics that buffer these problems, for example, the colour of asphalt can be changed to reduce evaporation and water loss, or sidewalks and roads could be made of cobblestones to improve drainage. Furthermore, groundwater could be recharged, or ecological corridors could be kept or built for species in each one of the roads (see Chap. 7). An interesting space of is the wetlands that comprise the Pantanos de Villa situated alongside the southbound highway on the outskirts of Lima (see Chap. 4). Upwelling from the phreatic zone here means that the highway has to be resurfaced frequently. Moreover, the wetlands have been diminished by soil compacting as part of nearby road and building construction.

Another general problem of pollution in the city are the heavy metals released by vehicles that reach the soil and pollute it. Specifically, in the case of Lima, heavy metals come from the mines in the water from the rivers and landslides (*huaicos*) or are transported from the mines awaiting exportation in Callao. The capacity of the soil to neutralize toxic substances varies and depends on several factors such as Cation Exchange Capacity, among others that is reinforced by the content of organic matter in the soil. Sandy soils typical in Lima have little capacity to absorb metals and prevent them from reaching the groundwater. One of the heavy metals from mining trails hazardous to health that can be found in the Rimac river basin is arsenic. However, water is constantly controlled to detect the presence of this metal but there is a potential risk which makes this constant control essential (SUNASS 2004: 114), especially in times of floods. Lead that comes from mining in the highlands is transported to Lima and stored in containers before it is exported, polluting soil and air with residue. Furthermore, stations sell petrol oil which is a source of contamination by lead and polycyclic aromatic hydrocarbon compounds, highly toxic and carcinogenic. Soil, highly contaminated with heavy metal, can be treated at temperatures of 1200 °C. This immobilizes heavy metals and eliminates polycyclic aromatic carbonated compounds but destroys all life in the soil which can only be replaced with great difficulty.

Two soil contamination problems that have been little studied are cooking oil from restaurants and gasoline from gas stations spilled onto the ground. Of all Peru's solid urban waste in 2017, 52.5 percent was appropriately treated, while 89.88 of the population had access to a solid waste collection service (SINIA 2020). A considerable expanse of land is degraded by municipal solid waste; 2,370.9 hectares throughout Peru in 2019, and 191.7 hectares in the department of Lima (SINIA 2020). In 2018, 0.6 kg of municipal solid household waste per inhabitant per day were generated compared with 0.9 kg per inhabitant per day of waste overall. The difference comes down to industrial activity.

To prevent pollution there are a great number of treatments, for example, soil can be isolated by a coating to prevent toxic products from reaching the groundwater (Endlicher et al. 2012: 124–125). These forms of protection should be considered when planning constructions of buildings. Some metals like cadmium or copper prevent microbial activity. These metals can be found in industrial areas in Lima and in fungicides or molluscicides to fight disease and pests. The higher the content of organic carbon of chemical substances, the better the absorption (Fent 2003: 36). Thus, oils, petroleum and grease or pollutants in the soil are absorbed easily by organic matter or by clay, remaining immobilized for a long period of time.

In general, anthropic soils in the city have a high content of nitrogen. Biogeochemical cycles are often altered, especially due to the increase of phosphates and nitrates (Francis and Chadwick 2013: 87). In the city, domestic sewage is the main source of phosphates and nitrates which is not a problem with proper treatment.

Urban soil can be classified in three categories: original soil, anthropic soil modified with different mixings and constructed soil (Endlicher et al. 2012). Original soils in Lima have two typical textures based on their origin. We can find sandy ground from the desert and alluvial, colluvial gravel with clay-loam and sandy-loam soil. Some parts of Lima have occupied garbage dumps where parks have been built. The soil in this area is highly flammable due to its high content of gases produced by anaerobic decomposition; we can find ethylene as well as a high content of organic matter.

Soil that comes from rivers has a large content of stones that give porosity to the soil and lodge microorganisms. Although some plants do not grow well when there are too many stones, others grow more efficiently in a properly aired soil. Soil that comes from garbage dumps is very rich in nutrients due to its high content of organic matter that triggers elevated microbial activity. However, it often has other toxic components such as metals. Ports like Callao where heavy metals are often stored before being exported show an elevated level of soil contamination.

Excessive soil fertilisation not only modifies soil pH but also reduces flora and fauna in the soil affecting biological conditions, triggering rapid growth in the plant that becomes sensitive to pests and disease (Chap. 3). In plants fertilised excessively, blooming and fruit ripening is delayed, growth is accelerated, the distance between internodes increases as well as the competition for light (Wohlgemuth et al. 2019: 312). The plants stretch out without investing in strengthening tissue. Therefore, more pesticides and fungicides are required.

6.3 Air

Air pollution in the city is caused by sulphur oxide, nitrogen oxide, carbon oxide, ozone, volatile organic compounds (VOC), polycyclic aromatic hydrocarbon compounds and articulated material (Francis and Chadwick 2013: 63). These emissions are mostly caused by vehicles and to a lesser extent by industries. They have resulted in a great number of bronchial and pulmonary diseases in the city and affect the health of plants.

Nitrogen oxide is more damaging when temperatures are high because under these conditions it transforms in ozone. This effect is produced when oxygen is released from nitrogen dioxide due to sunlight; the released oxygen is used by the molecule of oxygen to form ozone. Tropospheric ozone constitutes a greenhouse gas (Endlicher et al. 2012: 81). Undoubtedly, this reaction occurs in Lima where levels of sunlight and temperature are high.

The effects of high concentration of ammonium in the air have consequences in plant growth when there is a deposition of ammonium in the soil as acid rain (Fent 2003: 10). The increase of nitrogen by this means causes the plant to demand more water and other nutrients from the soil; if they are lacking, the plant becomes less resistant to stress as well as to the attack of insects (Fent 2003: 10). In Europe there is a high level of nitrogen deposition which seems to be the reason pine forests have perished (Fent 2003). It is worth noting that excess of nitrogen affects microbial symbiosis which pines require. On the other hand, we can observe a vicious circle because with the increase of fertilisation with nitrogen the plant needs more water and becomes more sensitive to pests and disease, consequently, the need for fumigation increases, affecting the health of inhabitants in the city. Agriculture contributes to a great extent to increase nitrogen oxide emissions. However, with organic agriculture there is a reduction in these emissions, so it is important to consider this form of fertilisation in green areas in the cities.

Damage to health produced by particulate matter (PM) is physical and depends on the size of the particles, being PM 10 and PM 2,5 hazardous to health by lodging in the lungs. The origin of particles that conform particulate matter is varied. They can come from burning, tire friction on asphalt, or plant pollen, among others (Endlicher et al. 2012: 81). Asphalt constantly releases PM 10 due to tire friction and pollutes the atmosphere. European standards for particulate matter determine that PM 10 can only exceed limits 35 days a year (Endlicher et al. 2012: 84).

Whereas vertical mobilization of the air in the troposphere, from the point of emission upwards is fast (Fent 2003: 33), horizontal mobilization of chemical substances in the air depends on the wind which blows them from side to side. The coast has constant winds to and from the continent; such is the case of Lima. Therefore, horizontal mobilization of the air is high. Marine birds and mammals absorb a great quantity of toxic chemical products that reach the sea. Metabolism of these animals makes them an efficient means of transportation for pollutants. This explains why an accumulation of pollutants can be found in the poles (Fent 2003: 33). This type of pollution is very common on the Peruvian coast near the cities and it affects the quality of fishing.

Air pollution also affects plants and animals. Species association can also be affected (Gallagher et al. 2011 cit. Francis and Chadwick 2013). Trees fulfil an important role cleaning the air by absorbing carbon dioxide and other pollutants. However, we should be aware that absorption can be little. It has been estimated that for United States, city trees only absorb from 1 to 2% of emissions (Nowak and Carne 2002 cit. Endlicher et al. 2012: 253–254). We can conclude that trees help clean the air, but they are not the only solution. There should also be a decrease of gas emissions in the city (see Fig. 6.1).

According to the Ministry of Health, in 2017 the morbidity rate from acute respiratory

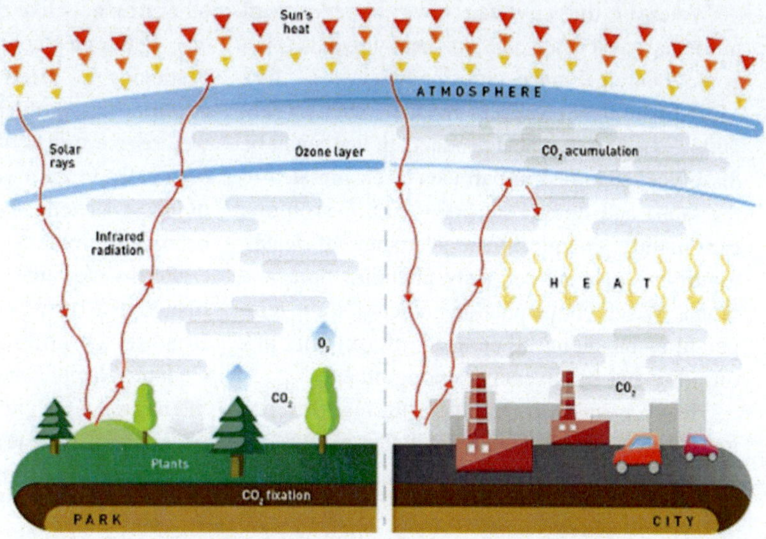

Fig. 6.1 Greenhouse effect. *Author* Ana Sabogal, designer: Juan Pablo Bruno

infections among children under five years of age was 87,494.11 per 100 thousand inhabitants; in Lima, the figure was 106,000 (SINIA 2020). The carbon dioxide equivalent emission rates are not kept up to date; the latest information is from 2012, and is not available at departmental level. Consumption of ozone-depleting substances, according to the Ministry of Production peaked dramatically in 2013, prompting regulatory measures that cut levels in half; however, levels began to creep back up thereafter. No data is available after 2017 or at departmental levels (SINIA 2020) (Fig. 6.2).

6.4 Water

The water cycle is altered in the city since urban density has an incidence on soil compaction. Therefore, permeability and flow of groundwater is modified (Francis and Chadwick 2013: 59). For this reason, for proper water management in the city it is essential to analyse the water cycle, its volume, how far the water comes from, canalization and changes in compaction in the water flow. In the city we can identify the following stages in the water cycle: evaporation, surface runoff, surface infiltration and deep infiltration (Arnold and Gibbons 1969 cit. Francis and Chadwick 2013: 60).

Water in the city can have an excess of nutrients and it often contains heavy metals (Hatt et al. 2004 cit. Francis and Chadwick 2013: 64). Most hydrolysis processes reduce toxicity (Fent 2003: 39). The chemical products in water can be dissolved, absorbed in particles or assimilated to the fauna and flora (Fent 2003: 31). Dissolution can modify chemical characteristics of water, modifying pH and seriously affecting flora and fauna. Water contamination indicators are temperature, biochemical demand of oxygen, pH, turbidity and biodiversity, among others.

Heavy metals can percolate through the soil and reach the groundwater (Endlicher et al. 2012: 105). From there, they can be carried to other spaces emerging from lakes or the sea. Microbial action transforms metals into organometallics, increasing their toxicity (Fent 2003: 37), since organometallics can be found in living organisms and bioaccumulate. Bioaccumulation produces biological magnification when the chemical product concentrates along the trophic chain affecting its different participants including humans. Such is the case of mercury, a cation which by microbial action transforms into methylmercury, a neurotoxin (Fent 2003: 39). Metal metalation is produced by enzymatic activities with lead, titanium, chrome, arsenic, zinc and selenium resulting in toxic compounds, except for arsenic whose toxicity decreases (Fent 2003: 39).

An excess of nutrients caused by sewages and emissions produced by fertilisers may trigger an increase in the Biochemical Demand of Oxygen as there are more consumers, the fecal coliforms, to decompose dead matter or they are fed by an excess of nitrogenated fertilisers. Among the most important water pollutants in the city we can find mineral oils and chlorinated hydrocarbons as well as pesticides, fungicides and nitrates that can reach the water and subsequently reach the groundwater (Endlicher et al. 2012: 105). Reusing the sediment produced in water treatment plants for sewage water can cause severe problems to soil due to accumulation of heavy metals, PCB residue or dioxins, additionally to other products like drug remains (Fent 2003:44–45). Few cities control the use of fertilisers and pesticides in green spaces. It is important to consider this problem when selecting an area as well as the origin of the soil for a public space, especially in the case of bio-orchards or instead of being beneficial to public health they could become a problem.

Peru-wide, the percentage of wastewater treated in 2017 was 89.4% (Instituto Nacional de Estadística e Informática—Dirección Técnica de Cuentas Nacionales cit. SINIA 2020). However, the figure varies greatly by location; for example, 98.97% of wastewater was treated in the department of Lima in 2018, while for the department of Ucayali in the Peruvian Amazonía, the figure was just 2.39% (Superintendencia Nacional de Servicios de Saneamiento—Dirección de Fiscalización cit. SINIA).

Biodiversity often diminishes in polluted places as few organisms are able to survive under these conditions. However, it becomes easy for new resistant species to install. An organism indicator of water pollution is the *Dreissena polymorpha* that feeds by absorbing and filtering water particles which it accumulates in its tissues (Minier et al. 2006 cit. Endlicher et al. 2012: 92). This species is invasive, distributed almost entirely around the world by displacing others due to its resistance. Other bio-indicators can be enzymes released as a reaction to pollution (Endlicher et al. 2012). Timely monitoring of substance is essential in order to prevent loss and damage to biodiversity.

Eutrophication is produced when there is an excess of nitrates and phosphates, the two fertilisers most often used in green areas. When these are used in large amount, they reach groundwater through percolation and displace to other sources of water. The excess of nutrients (nitrates and phosphates) results in an increase of floating plants and algae on the surface preventing the entrance of light, thus, algae cannot develop in the space immediately below and perishes together with aquatic plants. This increases the number of decomposers in the water, thus, the availability of dissolved oxygen decreases, which has repercussions on fish and other aerobic species which also perish, triggering an even higher demand of oxygen by decomposing bacteria. Finally, the aquatic ecosystem collapses. Therefore, an important parameter to determine water quality is the Biochemical Demand of Oxygen (BDO). When there is eutrophication, BDO increases dramatically. To a great extent, the city contributes with the eutrophication of water sources due to the indiscriminate use of fertilisers, pesticides and the disposal of fecal coliforms. On the other hand, towns or villages do not have an adequate water treatment for sewage that can cause eutrophication of water sources.

An excess of nitrates in the water also carries a risk to form nitrosamine in the organism during digestion which causes stomach cancer. There is also a direct relationship between the content of nitrate in the water and bladder cancer (Fent 2003). Nitrate on the water surface produces ammonium (NH_3) and nitrite (NO_2). Ammonium causes the death of surface fish since it damages the gills; on the other hand, nitrite has repercussions on the transportation of oxygen since it produces methaemoglobin (Fent 2003: 10).

Groundwater constitutes a very peculiar and specialized ecosystem, housing fungus, bacteria, worms and crustaceans which feed on organic remains transported in the water (Endlicher et al. 2012: 105), (see Fig. 6.2). These small organisms clear the pores in the soil and eliminate organic remains from groundwater keeping the underground water clean (Endlicher et al. 2012: 105). In Lima, there should be an analysis and monitoring of groundwater and the natural spaces that depend on underground water such as the Villa Marshes as well as the water that comes from evaporation like the "lomas" (hills). This will enable better water management in parks and gardens to reduce current loss by evaporation, transpiration, groundwater compaction and runoff improving efficiency of water use in Lima's desertic ecosystem.

The phreatic zone on Peru's coast is overexploited, as shown by national figures; according to the National Superintendency of Sanitation Services, 24.78 of water nationwide came from underground sources. No data is available for the city of Lima (SINIA 2020), but the level there is thought to be much higher.

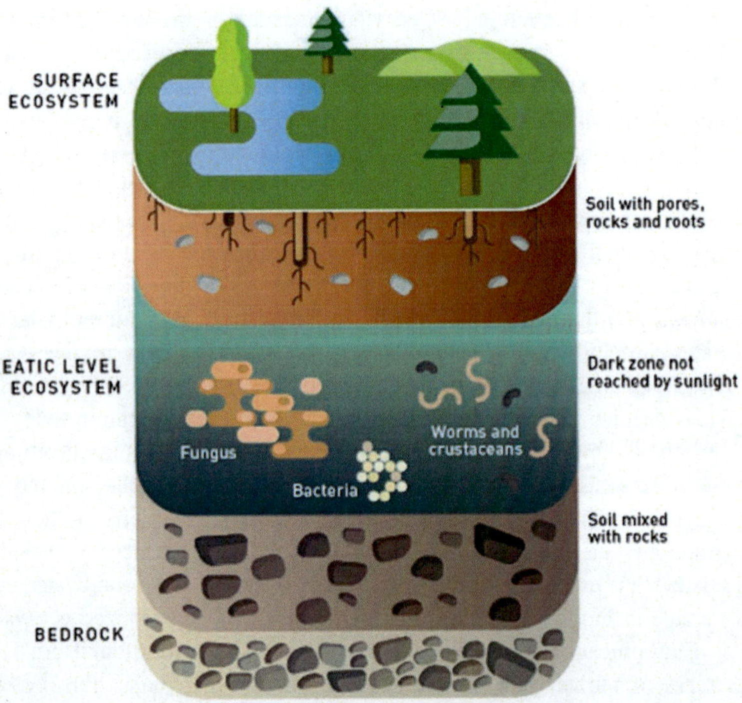

Fig. 6.2 Groundwater Ecosystem. *Author* Ana Sabogal, designer: Juan Pablo Bruno

6.5 Climate

In a city, environmental factors are altered; temperature is higher whereas humidity and wind decrease (Endlicher et al. 2012: 63). Longwave emissions in the city are 25% more than in the surrounding space (Kuttler 1985 cit. Endlicher et al. 2012: 64). It is estimated that cities release more than 70% of greenhouse gases worldwide (International Energy Agency 2008 cit. Bulkeley et al. 2012). Hence, the city is like an island of heat which worsens in summer at night due to nocturnal light and heat retention by buildings (Arnfield 2003 cit. Endlicher et al. 2012). Increase of temperature in the city is determined by city size, urban density, proximity to sources of water and amount of vegetation (Collier 2006 cit. Francis and Chadwick 2013: 58). In the city centre temperature can be up to 10 °C higher than in areas that are not urban (Zipperer et al. 1997 cit. Francis and Chadwick 2013: 58). In the case of Berlin increase in temperature that forms an island of heat has been estimated to be 3 a 4 °C (Endlicher and Laufer 2003 cit. Endlicher et al. 2012: 66). Green areas contribute to buffer increase of temperature in the city and mainly the trees, absorb carbon dioxide (Bulkeley et al. 2012). In coastal areas like Lima marine breeze in the evening buffers the increase of temperature in the city (see Fig. 6.3).

It is important to note that climatic factors vary depending on the height of buildings which means that climatic factors will not be the same on the ground floor as on the tenth floor. Furthermore, the third dimension should be considered (Endlicher et al. 2012: 72). A building's height will vary the projected shade and consequently the heat (Jendritzky, 1991 cit. Endlicher et al. 2012: 73). An object's dimension and volume will determine absorption and consequently, heat emission. Humidity is another factor that defines temperature variation between the day and the night as well as the hills and the ocean. There is more humidity at a lower height. Therefore, top floors in buildings and rooftops are less humid which should be considered when designing green rooftops.

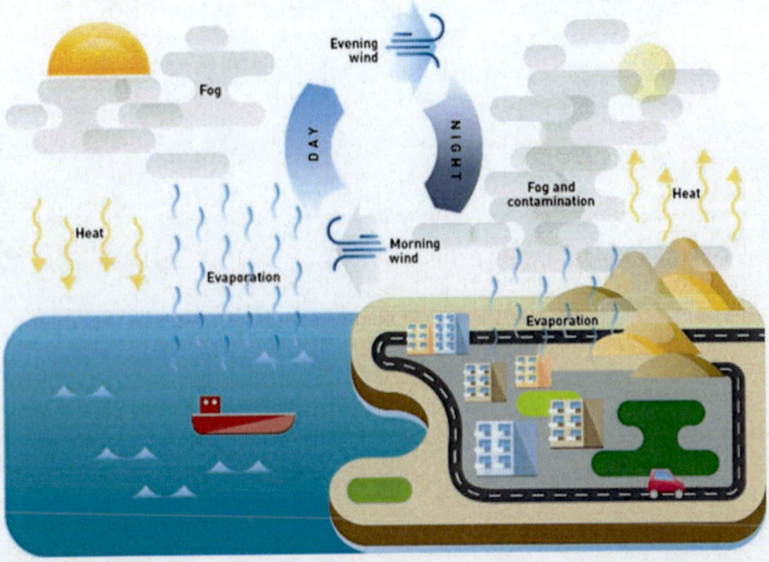

Fig. 6.3 Air flow during the day from and to the sea in Lima city. *Author* Ana Sabogal, designer: Juan Pablo Bruno

Lima has low building density (except for the centre), partly due to atmospheric humidity in excess of eighty percent for much of the year. In the parts of Lima closest to the sea, relative air humidity can go beyond ninety percent; in turn, western Lima is buffered by hills that trap atmospheric pollution, causing heightened levels there. Finally, air circulation close to the sea is high, reducing environmental concentration there but increasing humidity (see Fig. 6.3).

References

Bulkeley H, Castan Broto V, Edwards G (2012) Towards low carbon urbanism "from Local Environment". In: Wheeler S, Beatley T (eds)(2014) The sustainable urban development reader, 3rd edn. Routledge, London, New York, pp 101–106

Endlicher W, Kraas F, Krajewski Ch (2012) Einführung in die Stadökologie. UTB, Stuttgart, p 272

Fent K (2003) Ökotoxikologie, 2en edn. Thieme, Stuttgart, p 332

Francis R, Chadwick M (2013) Urban ecosystems: understanding the human environment. Routledge, USA, p 220

SINIA (2020) Sistema Nacional de Información Ambiental. https://sinia.minam.gob.pe/indicadores/listado Revised: 3/08/20

SUNASS (2004) Superintendencia Nacional de Servicios de Saneamiento. SUNASS, JICA. Lima, La calidad del agua potable en el Perú, p 259

Vercelloni M, Vercelloni V (2010) Geschichte der Gartenkultur von der Antike bis heute. WBG, Darmstad, p 275

Whiston A (1984) City and Nature, from the granite garden: Urban Nature and Human Design In: Wheeler S, Beatley T (eds) (2014) The sustainable urban development, 3rd edn. Routledge, London and New York, pp 61–65

Wohlgemuth T, Jentsch A, Seidl R (eds) (2019) Störungsökologie. Utb Haupt Verlag, Gernany, p 396

7 Urban Ecology

Abstract

This chapter proposes urban ecology as a science that studies city ecosystems. It develops a historical account of this concept and the different approaches that derive, comparing them and linking the concept to sustainable city and landscape ecology. It describes the characteristics, distribution and components of urban ecosystems. It centres on the analysis of biodiversity and species distribution in to the ecosystems. It describes the importance of ecological corridors for the distribution of the metapopulation and species biodiversity ensuring the adequate operation of ecosystems. It links landscape concepts with the basic principles of ecology and sets guidelines for the design of ecological corridors. It analyses disturbances as part of urban ecosystems as well as the response to these to achieve a balance. It describes ecological succession and the development and regeneration of ecosystems. Concepts are intertwined with specie classification based on origin and botanical characteristics, analyzing its repercussions on urban ecosystems and linking the description with examples of species for Lima city. The chapter revised the concept of urban ecosystem, analyzed the ecosystem of Lima city and proposed the way to enrich a livable city. The chapter links the Sustainable development Goal 11 with the goals 3, good health and well-being for people, and 7, clean energy.

Keywords

Urban ecology · Urban ecosystem · Ecological corridor · Ecological succession · Species

This chapter proposes urban ecology as a science that studies city ecosystems. It develops a historical account of this concept and the different approaches that derive, comparing them and linking the concept to sustainable city and landscape ecology. It describes the characteristics, distribution and components of urban ecosystems. It centres on the analysis of biodiversity and species distribution in to the ecosystems. It describes the importance of ecological corridors for the distribution of the metapopulation and species biodiversity ensuring the adequate operation of ecosystems. It links landscape concepts with the basic principles of ecology and sets guidelines for the design of ecological corridors. It analyses disturbances as part of urban ecosystems as well as the response to these to achieve a balance. It describes ecological succession and the development and regeneration of ecosystems. Concepts are intertwined with specie classification based on origin and botanical characteristics, analyzing its repercussions on urban ecosystems and linking the description with examples of species for Lima city. The chapter revised the concept of urban ecosystem, analyzed the ecosystem of Lima city and proposed the way to enrich a livable city. The

A. Sabogal, *Urban Ecology*, Sustainable Development Goals Series,
https://doi.org/10.1007/978-3-030-69905-5_7

chapter links the Sustainable development Goal 11 with the goals 3, good health and well-being for people, and 7, clean energy.

7.1 Urban Ecology

Urban ecology is the science that studies city ecosystems. The study of the city is relatively new. It has no more than one century and began with the great growth of metropolises. Urban ecology studies anthropic ecosystems, placing human beings as key participants of the ecology in the city. Chicago school defines urban ecology for the first time in 1925; at the time the definition only included the social dimension (Endlicher et al. 2012). However, the German school focuses on the study of the plants and ecological succession in urban ecosystems. The truth is that the city constitutes a cultural system where nature finds its space and develops in a new, peculiar way adequate for human beings.

Urban Ecology studies city ecology using ecological methods. This science begins with the conservation of ecosystems existing before the creation of the cities and with their restoration. It picks up principals from ecology science and it describes the landscape subdividing it in the elements typical of landscape ecology such as mosaic, patches, habitats, borders, fragmentation and ecological corridors. Urban ecology includes the concept of time to the analysis when it includes the ecological concepts of succession, resilience, balance and stability and when it analyses the ecological processes (Beatley 2011: 193). Urban ecology considers that humans are an element of the ecosystem and play the role of dominant species (Beatley 2011: 193). The objective of Urban Ecology is to preserve and restore ecosystems existing before the city appeared (Steiner 2011: 190). However, it is impossible to restore the ecosystems exactly to what they were before the city. This research adopts Urban Ecology in the sense that it considers urban ecosystems as systems which differ from their initial state, incorporating changes and adapting to the new conditions, which include both native and introduced species, forming new and dynamic structures different from the original ecosystems.

Urban ecology, urban landscaping, biophilia and ecological engineering are four different approaches to nature in the city from where the concept of urban ecological landscaping is derived. Whereas Urban Ecology derives from the conservation of pre-existing ecosystems and their restoration, for **Urban Landscaping** green spaces are ornamental areas separated from the urban pattern. In this case, green areas are relegated, excluded and treated separately (Beatley 2011: 191). Urban Landscaping as proposed by Waldheim Charles et al. (1990–2000) suggests the use of axles or visuals in the design, where green spaces work as patches in the urban ecosystem. This design was used in Parc de la Villette, Paris, designed by Tschumi and in the planning for Nueva York city, designed by Fresh Kills (Beatley 2011: 191).

Biophilia is defined by Wilson in 1984 as the need of human beings to approach other forms of life. From this perspective it is essential that city inhabitants be in touch with city nature (Beatley 2011: 181). Nature purifies and cleans the city and should be the centre of design in urban spaces (Beatley 2011: 183). A biophilic city typically has natural spaces and biodiversity including natural elements in its design (Beatley 2011: 182). It is an idealistic approximation in which city design incorporates nature from the beginning to form part of the city. However, most cities have not been designed following this concept. They did not consider natural spaces in their design until the city densified and it was necessary for health reasons to think about sustainable environmental solutions. Such is the case of London, during the industrial revolution, when the first public parks appeared. Park Birkenheard was built in Liverpool in 1847, whereas in France Napoleon III in 1853, had the royal parks remodelled creating the Bois de Boulogne, or Bois de Vincennes in Paris. The biophilic design is very often utopic.

The concept of **Ecological Engineering** is based on engineering and affirms that in the recreation of an artificial environment like the city, technical solutions can be found to enable the development of natural dynamics in the ecosystems. It postulates that engineering can give solutions to the contradictions between the natural and the anthropic. It is about studying and proposing suitable engineering technics so that the city has an adequate energetic circuit in which the energy emissions match the energetic absorption, and these be incorporated into the system. Therefore, it is about energetic balance attained by incorporating the adequate amount of nature into the city. Here, the purpose of nature is to absorb the negative balance of energy produced by the city. This current has many critics since it uses nature to solve a problem created by the city, without considering a reduction in the use of energy, losing sight of the ethic vision essential in comprehensive vision of planet conservation. Eco-cities, where environmental engineering prevails, can become cities where engineering is more important than society losing their democratic nature to become purely technocratic cities (Hagan 2015: 97). Some interesting attempts to recreate these concepts have been carried out in Masdar, Abu Dhabi, where environmental technology has been implemented without considering social aspects; another example is Dangtan, in the south of China, a city built by the government concentrating on energetic balance so that carbon emissions are zero which means that emissions are equal to absorptions. Finally, Gwanggyo in Korea, we can find an ecological business centre where ecology goes in hand with economics and the latter eventually determines planification axles (Hagan 2015: 97–100). In the first example we can see it is about a balance of emissions and in the last example it is about an economic balance. Both ideas incorporate environmental costs in their estimations as well as the purpose of green spaces, therefore it is purely a technical approach that does not consider different social or ecological aspects.

Ecological urban landscaping, postulated by Frederick Steiner at the beginning of this century, links cultural, aesthetic and ecological concepts (Beatley 2011: 191). The tendency in this design is to recover abandoned spaces and convert them into parks, incorporating ecological dynamics for ecosystem recovery and ecological succession as well as cultural and historical aspects, restoring abandoned spaces and revaluating historical memories. These designs can be seen in Schöneberger Südgelände park, in Berlin, Germany, where part of a World War II abandoned railway is the axel of the park's design which incorporates the concept of secondary succession letting nature take control of the new re-naturalized ecosystem naturally. In Bochum, Renania region, Jahrhundrethalle park incorporates an old abandoned steel factory to the design enabling nature to occupy the space and turning the factory into a cultural space. In both cases nature runs its course and human beings only facilitate the process by allowing natural processes to work and protecting them from destruction.

Lima has 608 historical monuments from the colonial era, most of them located in the historic center, while Lima and Callao together possess 300 *huacas* (ONU Habitat 2015: 46). Attention must be paid to this historical heritage, even if it cannot be conserved in its entirety. Some *huacas* have been protected and today constitute public spaces. One example is the Mateo Salado archeological complex in the district of Lima; constructed by the Yschma culture in 1100 AD, it was subsequently incorporated by the Inca empire into the Capac Ñan trail. The design of these spaces, and their establishment as public spaces, contribute to a reappraisal of Inca culture. At present, some of Lima's *huacas* are also used as museums or for staging theatrical productions. However, there remains much to be done if they are to realize their potential as public spaces. The same is true of historic monuments; many have colonial-era gardens could be included in the restoration of mansions from that era to become important public or cultural spaces.

The concept of **Livable cities** is linked to the previously described concept and makes us reflect on what a dreamed city would look like. This concept combines utopia with cultural multiplicity and social heterogeneity (Calthorpe

1993: 121). It is necessary to integrate and not separate or decontextualize natural spaces to design city communities adequately. In order to ensure a good quality of life for inhabitants it is important that the city be suitable for pedestrians (Calthorpe 1993: 122). Housing communities should include nature in their design, preventing isolation between natural and constructed space. This current proposes that the design of public spaces and green strips are a good way to integrate society (Calthorpe 1993: 122). This also enables the development of pedestrian spaces as well as bike lanes to contribute with the reduction of emissions caused by motor vehicles creating a space where the population approaches nature.

Finally, **Reconciliation ecology** proposes the need to have knowledge of the ecology in the city and reconcile it with city dynamics to make it work. Furthermore, ecologies, natural dynamics and anthropic alterations should reconcile through the promotion of new habitats by facilitating the diffusion of species with the use of key species and other ecological processes to achieve it. Nature reconciliates with the citizens and finds its space in the city. The Table 7.1 systemised the ecological concepts used in the landscape design described.

This is an especially interesting challenge for Lima and its sustainable development, but there is everything still to do. It should be noted that the Plan Metropolitano de Desarrollo Urbano al 2035—Municipalidad Metropolitana de Lima seeks to prioritize sustainable urban transportation systems for pedestrians and cyclists (ONU Habitat 2015: 23). Also worthy of note is the reappraisal of archaeological zones and the conservation of hills and environmental corridors through recovery of the ecological structure of the Chillón, Rímac, and Lurín rivers. ONU Habitat (2015: 26). Successful realization would certainly make a contribution to ecological reconciliation, and to the sustainable development of the city.

Accomplishing a sustainable city depends on many factors. Examples of sustainable cities are Destiny in Florida, USA, Hamburg in Germany and Auroville in India or Lilypad in Belgium. However, all these cities are artificial from the point of view of design, where the government fulfils an important role (Hagan 2015: 100–116). These cities are small, so many authors affirm that livable cities have a size limit (Mumford, 1996 cit. Hagan 2015: 120). Geddes considers that cities should be small and interconnected (Hagan 2015: 119). The study made by the European Union reveals that mid-sized cities have a better economic position than big cities (Giffinger et al. 2010 cit. Hagan 2015: 122). This was the basis for the proposal to create interconnections between the small industrial cities in Rhineland, creating an interconnected railway network. The region has 11 interconnected cities with around 5 million urban inhabitants. Last century it was one of the most contaminated regions due to coal and steel mining; currently after the restoration,

Table 7.1 Ecological concepts in landscape design, own authorship

Ecological concepts in landscape design	
Concept	Distinctive characteristic
Urban ecology	To preserve and restore natural ecosystems
Urban landscaping	Natural elements constitute patches in the urban pattern
Biophilia	To incorporate natural elements and ecosystems to the city's design
Ecological engineering	To use technical solutions in order to achieve energetic balance in the city
Ecological urban landscaping	To define ecosystems as cultural and historical spaces, recovering and re-naturalizing natural ecosystems
Livable city	Design of natural spaces implies a vision from culture and society
Reconciliation ecology	Urban ecosystems are new ecosystems, reconciled with the city

decontamination and ecological recovery of the Rhine river, it is still a relevant industrial area as well as a touristic centre.

Moreover, and as mentioned elsewhere, the Plan Metropolitano de Desarrollo Urbano al 2035—Municipalidad Metropolitana de Lima also proposes the development of a polycentyric city made up of 58 centralities (ONU Habitat 2015: 43). This would enable a reduction in commuting and thus in air pollution, leading to improvements in the population's health.

7.2 Urban Ecosystem: Biodiversity Periphery/City Centre

Biodiversity in urban ecosystems can be richer than that of adjacent agricultural systems. This happens for many reasons among which the most important are the existence of monocrops, the use of agrochemicals, the characteristics of agricultural ecosystems. On the other hand, the lack of predators in the city.

Urban patterns and population density are essential in the distribution of biodiversity. The density of urbanizations causes habitat fragmentation and has a direct effect on biodiversity (Francis and Chadwick 2013: 67). Characteristics such as time of urbanization and distance from the city centre determine the patterns of biodiversity (Pauchard et al. 2006 cit. Francis and Chadwick 2013: 69). Older urban spaces have more biodiversity because more time has elapsed since the species were introduced (Kowarik 1990 and Sukopp 1998, cit. Francis and Chadwick 2013: 68); in older spaces, ecological succession has had more time to enable the development of the ecosystem and the trophic chain which reflects on an increase in biodiversity.

Biodiversity in great parks is without a doubt larger than in smaller parks as there is a larger habitat (LaPaix and Freedman, 2010 cit. Francis and Chadwick 2013: 75). A city has an endless number of ecosystems and very diverse ecological niches. Different type of ecosystems will be occupied by different species; this will depend mainly on the characteristics offered by the habitats. The predominant species in the city are generalists as well as introduced and invasive species (Francis and Chadwick 2013: 115). The bigger the space the more specialist species there will be (Francis and Chadwick 2013: 109). This, however, will depend on amount of border occupied by the ecosystem. It is important to note that borders privilege generalist species. Therefore, bigger parks have more habitats where specialists can develop. Urban habitats are different from the original habitats of the species; even native species from the area before the construction of the city do not find their original habitat. That is why the distinction between native and introduced species is not applicable in these spaces where species are anthropic, like pigeons, cats, or rats found in every city in the world that have evolved together with the city. These species have lived with human beings for centuries and without them they would struggle to prosper. Most of the benefits would be reaped in Lima, where most of the green spaces host ecosystems of introduced species whose survival depends on irrigation.

Most plants are brought from other spaces and need to adapt to an urban ecosystem in which they should prosper. Urban ecosystems constitute new vegetation associations developing new ecotypes and adaptations where the species tolerate high levels of pollution and new microclimates and form new associations (Francis and Chadwick 2013: 15). There is great biodiversity in the city because of the introduction of species (Galluzzi et al. 2010 cit. Francis and Chadwick 2013: 17). In a city an ecosystem needs to be built; the first to arrive are the introduced species planted by humans; generalists reach these new ecosystems occupying empty niches and initiating the ecosystems; the specialists then arrive when the ecosystem re-builds provided that, as often happens in the city, the ecosystem is not modified. Furthermore, there are many anthropic spaces such as apartments with or without plants, balconies, external walls, green tops and other spaces that also constitute ecosystems in the city. Thus, urban ecosystems are diverse and include a diversity of species.

In the city we can find a great number of new distinctive habitants. Most of them are micro habitats with generalist species that conquer their space, with exceptions (Francis and Chadwick 2013: 93). For example, habitats that offer a wall are very diverse, with different temperatures, humidity, sun exposure based on height (Francis and Chadwick 2013: 92). Also roof tops can lodge a variety of birds (Francis and Chadwick 2013: 95). The number of invertebrates in the city is very large and it is even larger in gardens with arboreal species with abundant green spaces (Smith et al. 2006 cit. Francis and Chadwick 2013: 77). Spaces like corners or spaces between tiles are interesting for the biodiversity that surges in the shadows and humidity of these corners (Kowarik 2003). Many species of birds and insects can only be found in the cities since these species have developed for centuries in them (Endlicher et al. 2012).

There is fauna in different city environments such as urban zones, industrial zones, houses, parks, and gardens. There is different fauna in closets, balconies, storage room and flowerpots. Houses also lodge species like lichens, fungus or insects. For example, stone covering a wall can have lichens, fungus or nests with small insects or spiders that form a very special, peculiar but small ecosystem (see Fig. 7.1). Each one of these ecosystems conform different communities with specific fauna. In Lima, high atmospheric humidity has led to a proliferation of fungi, mildew and dust mites within and outside dwellings.

Plants and animals in the city constitute a metapopulation that covers different parts of the city and the surrounding field. There is communication between individuals that conform the metapopulation within and outside the city and with the adjacent cities (Endlicher et al. 2012). Especially pedestrian lanes, but also highways, can be dispersion routes for species, particularly generalists contributing to genetic recombination and consequently the uniformization of the population biodiversity.

Zones like route intersections can be isolated in central zones and form ecological islands (Francis and Chadwick 2013: 100), where their own species can develop. City habitats can be found in mosaics (Endlicher et al. 2012: 143).

Fig. 7.1 External green wall, Berlin. *Author* Ana Sabogal

Small vegetation mosaics have high levels of vegetation. However, the fauna of birds is highly sensitive to fragmentation (Forman 1996 cit. Francis and Chadwick 2013: 68–69). A distinctive example of an island of vegetation was the area in between the two Berlin walls. In this zone, there was a complete ecosystem where species developed their own dynamic and even foxes could be found. The Fig. 7.2 show the different Urban Ecosystems. Certain spaces in the centre of the city, such as viceregal cloisters or convents, harbour species in isolation that have developed dynamics all of their own.

The form and structure of vegetation defines the space covered by the ecosystem and influences the physical distribution of plants and animals. An ecosystem's physical structure only includes bidimensional space, determined by the height of vegetation. Thus, there is a direct relationship between plant size and species diversity. The higher the arboreal structure, the more niches there will be, thus, the more diversity there will be. However, the space it occupies

Fig. 7.2 Urban Ecosystems. *Author* Ana Sabogal, designer: Juan Pablo Bruno

in the ecosystem is tri-dimensional, formed by length, width and height of the species, including the depth of the roots. Therefore, Biological Structure considers biodiversity, and it is calculated taking into consideration dominance (number of individual of each species), relative abundance (number of individuals of each species in relationship to the total number of individuals of all species) and the diversity of species (total number of species). The parameters that define the Biological Structure are related to the diversity of niches originated by space diversity. In this way, Physical Structure facilitates Biological Structure. This means that animals that live in green areas depend on the vegetation offered and on the ecological niches that can be found. Furthermore, arboreal species promote diversity of birds that nest in the trees and the dominance of grass promotes a different species of bird like the groove-billed ani, and earth worms, or insects, whereas rats prefer areas with bushes such as leafy and dense vines. The following list systemised the characteristics of the urban ecosystems. In Lima, the trees host large numbers of birds that feed on insects. Species may be native, such as the long-tailed mockingbird (*Mimus longicaudatus*) and the cinereous conebill (*Conirostrum cinereum*); or non-native, such as the saffron finch (*Sicalis flaveola*) and the blue-grey ranger (*Thraupis episcopus*), indigenous to the Amazon basin, which have rewilded.

Urban ecosystem characteristics

- High presence of generalists
- High biodiversity
- Fewer predators
- Dominance of introduced species
- Species resistant to pollution
- Vegetation Mosaics
- Ecological corridors.

Some spaces in the city offer plants and animals similar conditions to their natural habitats. For example, mites use bird nests in their natural habitats whereas in the city they live in similar habitats since houses keep warm temperatures and offer a protected space; mites lodge in rugs, and furniture causing allergies to human beings (Francis and Chadwick 2013: 122). The process of adaptation in the city is continuous and depends to a great extent on the genetic characteristics of a group of individuals of a species that arrives and adapts to the new conditions in the city, also known as the "founder effect" (Francis and Chadwick 2013: 126). Thereby, the individuals that arrive first bring specific genes that will prevail among the population of this species in the city where the population is isolated constituting an endogamic genetic group. This adaptation often implies a modal selection, with genes that in nature do not have the necessary conditions to subsist natural selection but find an appropriate habitat in the city. At that point, there is a genetic selection of genes that in wild populations would appear in fewer number or would not appear at all. The city is then a genetic island. Populations in the city are isolated and there are not many individuals of introduced species. Therefore, there is a genetic derivation in relationship with wild populations. These populations distance themselves from the original population and become more dependent on human beings. Therefore, they become a different population separate from the wild populations, eventually becoming anthropic species, provided that their wild relatives do not win the competition outside the ecosystems of the cities.

Like plants, many animals might prefer the city to the countryside since it is easier to find food and there are fewer predators. Furthermore, temperature is higher, and winds are moderated by buildings and other constructions. Thereby, in the city, species sensitive to predators can develop without being hunted or consumed as would happen in their natural habitats. The existence of both native and introduced species means that Lima's fauna is very diverse, often more so than the country's natural ecosystems. Fauna in the city includes native, introduced as well as adapted species. Therefore, it is very diverse and often more varied than in their natural ecosystems. Some domestic animals have become part of the wild fauna of the city. For example, the pacific parrotlet (*Forpus coelesteris*) indigenous from the dry forest has spread in the parks of Lima and become wild again together with many cats. In the last years some new species from other areas have appeared in Lima such as the saffron finch (*Sicalis flaveola*) or the blue-grey tanager (*Thraupis episcopus*), indigenous from the amazon basin, which were initially introduced as pets for their beautiful yellow and blue colours respectively, and later spread throughout the city. However, birds in the city must face multiple dangers such as traffic and windows in buildings (Endlicher et al. 2012). Furthermore, their predators in Lima are the cats (*Felis catus*) that hunt them as well as the Guayaquil squirrel (*Sciurus stramineus*) that eat their eggs. Birds are also threatened by the pesticides and insecticides used in the parks. All these circumstances modify their behaviour and habitat. It has been noted that birds in the city begin singing earlier and finish later, using artificial light. (Endlicher et al. 2012: 148). After these adaptations it would be difficult or impossible for these species to return to their original habitat.

The heterogeneity of habitats in the city enables the diversity of fauna in the metropolis (Francis and Chadwick 2013: 158). So, it is interesting to study how they originate and form diverse ecosystems in parks. Furthermore, through adequate design it is possible to increase the diversity of species occupying parks by conditioning them so that species can find their habitat. For example, planting nut trees like pecans will attract squirrels and parrots, or by placing wooden nests in parks there will be more birds. In Lima, the cultivation of nut trees such as pecan attracts Guayaquil squirrels (*Sciurus stramineus*) and parrots such as the red-masked parakeet (*Psittacara erythrogenys*), while water bodies entice species such as the Rufous-collared sparrow (*Zonotrichia capensis*). Habitat selection will depend on the species, some prefer small spaces with borders and others more sensitive only develop in ample spaces (see Fig. 7.3).

Habitats can be recreated to promote the development of species. For example, in the market there are a several nests or water fountains which can be installed in trees, balconies, or other green areas to promote bird nesting. Whereas trees promote the presence of parrots and squirrels, bushes promote the development of species like the blue-black grassquit (*Volatinia jacarina*) and grass attracts the groove-billed ani (*Crotophaga sulcirostris*). Each species occupies a different space; even in a tree each one has a height for nesting and occupying. Some position themselves in the middle part of the tree, others in the base and finally some prefer the top of the tree to be able to spot their prey from there. Thus, antennas or high buildings are often used by the

Fig. 7.3 Nests for birds placed in a park. *Author* Ana Sabogal

peregrine falcon (*Falco peregrinus*) in order to spot its prey easily. The decision of what species to prioritize will depend on the key species, since they will promote the development of other species. We also should consider the trophic chain that it promotes. For example, the peregrine falcon is excellent for the city because it eats rodents, small squirrels, rats or mice. Frogs which are appearing in Lima due to climate change feed on cockroaches.

7.3 Ecological Corridors

Urban ecological corridors are natural or designed spaces with vegetation that interconnect parks, green areas and other natural spaces in the city. Their role is to connect the population with species in order to prevent islands of vegetation. Ecological corridors unite patches of vegetation reducing fragmentation and enabling the connection between species so that metapopulations can function. Whereas ecological corridors connect species, fragmentation divides them.

Ecological corridors do not only unite land spaces but also unite superficial or subterraneous aquatic spaces as well as air spaces (Hagan 2015: 43). Corridors also enable inhabitants to walk and get in touch with green areas in the city. Therefore, it also contributes to inhabitants' health. It is important to consider the air corridors in the city in order to enrichen fauna with birds. It is common for land corridors in the city to be linear to make motorized and pedestrian transportation easier. However, when planning corridors, design is not enough. It is also necessary to consider and ensure the habitat of species in order to develop ecosystems. In this sense, more important than linear corridors is the network that connects them (Francis and Chadwick 2013: 164). To design the corridors, it is essential to consider the circuits, the habitat of species to be promoted along the corridor and connectivity with patches. Thus, the design of the corridor is complex.

Before defining the trajectory of the corridors and interconnecting the patches, each patch should be studied. It is important to note that all patches do not have the same quality. There are patches where reproduction exceeds death and vice versa, others that require constant recolonization by populations (Turner 2001). The size and quality of the patches is vital for the selection of species. There are species that require patches to have a minimum size (Francis and Chadwick 2013: 38). These are basically generalists that live in the city, whereas, wild specialist species require larger spaces. For this reason, the form of the park as well as the distance from the periphery to the centre influence the species that are present. Non-native species develop more easily in the borders of each patch (Hansen and Clevenger 2005 cit. Francis and Chadwick 2013: 29). When corridors are very small and narrow and interconnect very few patches there is a border effect, favouring and enabling the development of species tolerant to urban pressure (Francis and Chadwick 2013: 29). Wide corridors enable the development of more sensitive species that will place themselves inside the corridor. It is worth noting that most of the city's ecological corridors are narrow, which fosters fauna such as the West Peruvian dove and the Guayaquil squirrel, to the detriment of the kind of specialist species that are few in number in Lima, such the Peregrine falcon (*Falco peregrinus*); and of solitary, territorial birds like the scarlet flycatcher (*Pyrocephalus rubinus*).

There is a direct relationship between the increase of patch size and the number of species (Francis and Chadwick 2013: 38). The bigger the patch, the more species there will be. This relationship can be calculated with the following equation:

$$S = kAz$$

where:

- S is the number of species
- A is the area of the patch
- k and z are constants that vary based on the case, K represents the group of regional species (K) and z represents the species growth that depends of the relationship between species and area (z) and.

Bigger patches have a larger number of species with more stability where local extinction diminishes, and the availability of niches increases. In this way, there are less disturbances, so population of species become more stable and their possibility of speciation grows (Turner and Tjorve 2005; Spiller and Schoener 2009 cit. Francis and Chadwick 2013: 39).

To determine the distance required by species between patches in order to enable interconnection and form metapopulations, patch size as well as species abundance and density could be measured corelating it with the isolation between patches (Francis and Chadwick 2013: 41). Isolated patches only allow mobility of highly mobile species that disperse easily (Prevedello and Viera 2010 cit. Francis and Chadwick 2013: 39). We can distinguish between functional and spatial connections which are essential for the dispersion of each species (Francis and Chadwick 2013: 40). Spatial connections are land connections, whereas functional connections do not only include land connections but also air and water connections. The distance between the patches of green space that each bird species requires is highly variable, and their needs depend on the level of territoriality. Territorial species such as the peregrine falcon and the Amazilia hummingbird (*Amazilia amazilia*) require lower niche density, while their less territorial counterparts, such as the rufous-collared sparrow, are less wary of humans and find spaces with greater ease.

Speciation refers to the formation of new species at a speed which depends on the species complexity. In the first stage the species forms a separate genetic group, isolating itself from the other members of the metapopulation; subspecies are formed in this stage and start differentiating from each other in the second stage forming new species after many generations. The possibility of speciation is higher when interconnection between patches is lower and the population is more isolated. This depends not only on the patch size but also on the interrelationship between patches. In the city many populations are isolated without connections with other members of their species. It is important to note that these species will be very sensitive to changes so common in the cities. These species have to adapt to changes in the urban environment. An interesting example for Lima is the scarlet flycatcher *Pyrocephalus rubinus*); the bright red coloration of the males has gradually turned a reddish-brown for camouflage. Though it will not create a new species, this lenghty process of change can give rise to a distinct ecotype followed by a new group within the same species. The Amazilia hummingbird, found in Lima's parks, is another species to have undergone a slow speciation process, evolving a long beak and the rapid wing movement for which the bird is known.

The classic hypothesis of Intermediate Perturbation postulated by Cornnell (1978) cit. Wohlgemuth et al. 2019: 78 states that it is ideal to have certain levels of perturbation in the ecosystems in order to increase biodiversity. Subsequently, Hudson (1994) cit. Wohlgemuth et al. 2019: 41 adds to this hypotesis that an increase in biodiversity will only occur when there are resources available. This is correlated with the free niches in the ecosystem of the city.

These niches are constantly disturbed and renovated so diversity is changing not only because new species are planted but also due to the constant adequation of local species. Disturbance allows maximum biodiversity since competition of superior species is reduced, preventing dominance of species (Francis and Chadwick 2013: 41). Disturbance forces evolution and species adaptation. It is measured by the intensity, frequency and duration of the disturbance. For this reason, similar patches develop differently based on the degree of disturbance. This complexity makes it difficult to determine the size and length of the corridor (Francis and Chadwick 2013: 42–43) as well as the genetic characteristics of population group of each species. Some species adapt to disturbances, becoming able to live under these conditions. Growing on the banks of the Rímac and Chillón rivers are species such as Peruvian pepper (*Schinus molle*), Humboldt's willow (*Salix humboldtiana*) and Peruvian sauco (*Sambucus peruviana*), which adapt to fluctuations in the water level.

Rivers represent excellent corridors; they carry seeds and renew species; they are connected by superficial water and underground water as well. Therefore, they are very sensitive to pollution which is the case of the Rimac river and all rivers on the Peruvian coast. As all other rivers in Peru, Rimac river has a variable growth during the year. During flood season the river can grow several times its size. This season enables the development of ecological cycles and the renewal of nutrients in the ecosystem. Therefore, there is a regulation that determines a minimum distance of 300 m between the river edge and constructions. However, this regulation is not enforced so houses are exposed to the risk of floods every year. Furthermore, the river is exposed to contamination by the sewage of adjacent houses in spite of the regulation of the minimum distance calculated based on the river course.

In the design of corridors, it is ideal to include the regional scale (Francis and Chadwick 2013: 166). The use of tunnels and bridges as ecological corridors enable communication between fauna (Francis and Chadwick 2013: 154). It is important to take this into consideration when building highways and roads that make transit difficult for species. In case of corridors of aquatic ecosystems, connections between underground water and other aquatic ecosystems should also be considered whether they are basins or wetlands. In 1969, McHarg proposes the use of a bigger scale in the design of green spaces (Beatley 2011: 191). McHarg proposes and implements the restoration of the hydrologic corridor for Texas (Steiner 2011: 192). In his design, he analyses the physical and biological characteristics of the space (geology, hydrology, topography and vegetation) and subsequently, the socio-cultural characteristics using and analysing the maps of space (Hagan 2015: 63). The regional perspective can also be incorporated in this design, through space planning which implies the use of the environment and its resources (Hanna and Slocombe 2013: 30). This form of planning is often difficult to achieve since comprehensive planning is frequently seen mistakenly as a reduction of governmental power (Hanna and Slocombe 2013: 33). Effective planning integrates population and takes into consideration formal and informal relationships to make it realistic and accomplish a landscape design that is feasible and sustainable (Hanna and Slocombe 2013: 44–46). However, it is difficult to arrive to a consensus. It takes time, transparent processes and government continuity which is not always possible in Peru.

In the design of public spaces, in order to form and restore ecosystems, ecological communities should be mixed, formed by a diversity of species and spaces. Monocrops and a lack of species diversity leads to more pests, which in spaces without biological control like most parks in Lima result in an unrestrained use of pesticides. Although there are some municipal ordinances that regulate the usage of pesticides, municipalities that apply biological controls, and important initiatives such as the Ecological Agriculture Network, which promotes the avoidance of pesticides, most of these are aimed primarily at urban agriculture.

In design and restoration, the formation of islands should be avoided. Therefore, vegetation strips in the city can ensure connections

(Calthorpe 1993: 122). These strips can be continuous or formed by a network of overlapping ecological corridors. Ecological restoration should consider scales, contemplating the landscape borders and habitats (Francis and Chadwick 2013: 137).

The concept of green strip that surrounds the city and separates the urban space from the rural space enables the transition between these spaces (Amati, 2008 cit. Francis and Chadwick 2013: 151). This concept was developed first in Great Britain and then it expanded around Europe and United States in the twentieth century (Ignatieva et al. 2011 cit. Francis and Chadwick 2013: 151). Migge (1881–1935), during the pre-war develops the concept of a green strips outside the city of Berlin, in Leipzig. Migge includes the concept of public health and food security as explained in Sect. 4.3. However, this type of separation between urban and rural is currently questioned and has been replaced by interconnected ecological corridors (Francis and Chadwick 2013: 152). Subsequently, the concept of green strip was modified in 1985 and redefined by Behrens in 1985. The new concept was implemented in 2003 in the city of Frankfort am Main (Endlicher et al. 2012) enabling interconnections between species and space and promoting urban ecosystems. As mentioned previously, urban ecosystems can have more biodiversity than rural spaces. Therefore, green strips should lodge and give refuge to species in the city while in transition to rural spaces.

The concept of *Garden city* is more current. It was proposed by Ebenezer Howard in England in 1898, who planned a city with spaces for agricultures as belts of the industrial cities and intermediate cities interconnected with agricultural spaces; the city is developed in concentric circles with parks and groves interconnecting spaces and an external circle with an industrial area (Endlicher et al. 2012). Currently, to respond to the problems generated by the great metropolises, especially those related to health issues, the concept of intermediate cities has been proposed as an option for a more balanced deconcentration and interconnection. However, this possibility should be contemplated while planning the city so that it becomes an interesting option for migrant populations from the countryside to the city. It is also essential to contemplate the development of health centres as well as education centre. Deconcentration should be planned in order to control the growth of large metropolises.

7.4 Ecological Succession

Ecological succession describes changes that occur over time in an ecosystem. It refers to the ecological processes and changes along the time, in both urban and natural ecosystems. Succession occurs in different stages. When an ecosystem goes through these stages and has developed the conditions to establish, we are referring to a primary succession. The first stage of succession is the pioneer stage when species arrive to an area devoid of life and need to survive to give place to the development of an ecosystem. The next stages are called Seral which can be several. At this point, species reach the ecosystem and occupy it slowly. When there is no more free space for occupation and migration is only possible at the expense of another installed species, the ecosystem has reached a balance called Climax. In his stage, everything is useful. Plants produce food for herbivorous that feed carnivorous completing the trophic pyramid. Provided there are no new species or conditions do not change, the ecosystem remains stable. If changes occur, the ecosystem can go through a stage of Post-climax and subsequently perish or readjust and develop new conditions with a new distribution of roles for the species conforming the ecosystem. New species can incorporate and occupy abandoned ecological niches or conquer them by displacing species (see Table 7.2 and Fig. 7.2).

Secondary succession takes place when the process of succession is disturbed in any of its stage, but it is able to re-build with new dynamics, roles and species. The interruption in the process of succession that generates the secondary succession can be identified when there are remains of vegetation from previous stages, burnt trunks, wetland remains, species from

Table 7.2 Succession stages and characteristics, own ownership

Stages of succession and characteristics		
Primary succession	Secondary succession	Characteristics
Pioneer stage	Disturbance	Ecosystem begins
Seral (s)	Seral (s)	Plant dominance and generalists
Climax	Climax	Maximum biodiversity, ecosystem stability and balance. Complete trophic chain
Post-climax	Post-climax	Dominance of some species, decrease of biodiversity

different stages, or native plants that survive in city ecosystems. Species brought from different parts of the world and planted in the city form new ecosystems in which we can find some species from the original ecosystem. It is interesting to see how even though human beings choose what species are planted as well as the conditions of space, once the park is developed, local species integrate the ecosystem, find refuge and develop new roles. Human intervention will regulate the constant changes and readjustment of the ecosystem by fertilising, pruning and other activities common in urban horticulture. Due to these constant disturbances, the species installed are tolerant to these disturbances. Disturbance promotes species diversity and enables new species to install (Wohlgemuth et al. 2019: 14). New species go through the adaptation processes typical of the species. However, they also interact with other species native of the ecosystem, becoming part of a community and being functional in the new ecosystem which human beings conform. This process in which the ecosystem forms and changes occurs in stages and is part of urban ecological succession (see Table 7.2).

Secondary succession dynamics like the ones found in the city involve patch dynamics where, in principal, each patch is equally important since they are interconnected and lodge the dynamics of a diversity of genes that conform each species. On the other hand, islands of vegetation lodge rare species that might be important to reestablish the ecosystem after changes or disturbances. The facilitation hypothesis is applicable in urban ecosystems which means that in a succession some species facilitate the establishment of others by forming a habitat that is appropriate for them. In the city, this is the case for introduced species that are facilitators for native species since introduced species are not in their original ecosystem so they cannot compete.

In the city there is a constant and dynamic introduction of species. Limiting factors can determine the route of succession and the communities that install (Francis and Chadwick 2013: 120). Ecological communities in the city are different from the ones in the countryside. According to Meurk 2010 cit. Francis and Chadwick (2013: 118), we can identify three typical communities in the city: remnant communities similar to the ones in the original communities, spontaneous communities that find a city substratum to colonize and develop and those deliberately designed and planted by human beings. Whereas the promotion of remnant communities entails conservation, the creation of deliberate spaces implies the creation of new ecosystems over the conservation of native ones.

In their battle to install, exogenous species can be invasive and displace local ones. However, they can also facilitate space for the local species that are being displaced by the changing conditions of the city.

In the city, natural regeneration of species is from an ecological point of view an interesting form of conservation since it ensures the survival of native species. Spaces like the Villa Marshes wildlife reserve situated in the south part of Lima city, are important for local species or they would have to compete with introduced species. Overall in these spaces it is important to preserve diversity of the birds nesting in the cattails (*Typha sp.*). These birds are often migrating, and the wetlands are essential to their survival since they find refuge when it is winter in their native land. They feed on the species that live on the upwelling characteristic of wetlands and form lagoons surrounded by cattail. Furthermore, the reserve is connected by underground water with the rest of the wetland along the coast of Peru. Many of them suffer the pressure of the cities which is the case of the Villa Marshes or the Trujillo wetlands. The pressure generated by real estate compacts the soil and prevents the adequate flow of water whereas groundwater is contaminated by sewages. All this alters the ecosystem and causes a change in their distribution of species. The new ecosystem must adapt; exotic species can occupy the spaces left by the native species and form a new vegetation composition with native species coexisting with introduced species and often thanks to them. The role of cattail is essential since it filters the water and enables birds to nest. Wetlands are important because they are interconnected along the Peruvian coast and they are part of the bird migration route from both the highlands as well as North America. Without them bird diversity would be affected.

7.5 Species Classification

Depending on their form of interaction in the ecosystem and their vegetation community, species can fulfil their distinctive roles. Species have three differential capacities that allow their classification: their persistence after a disturbance, the forms of growth within their vegetation community and their cycle of life (Noble and Slatyer 1980 cit. Wohlgemuth et al. 2019: 81).The mentioned capacities will determine their role, permanence and dominance in the ecosystem.

Urban species can also be classified in relation to their preference for urban ecosystems in three groups: those which do not like the city (urbanophobas), those without a preference, developing well in both the city and countryside (urbanoneutral) and those with love for the city (urbanophilas) (Wittig 1998 cit. Endlicher et al. 2012: 131). The next section describes the roles of each group of species and classifies them based on it.

7.5.1 Dominant Species

A dominant species is the one that determines the roles of others in the ecosystem. They are very tolerant to competition and displace others occupying a privileged place of access to resources. Dominant species are not always the same ones. It will depend on the composition of the ecosystem. A species that is dominant in one might not be dominant in another with a different composition.

A disturbance can cause a variation from one dominant species to another. This means that the dominant species may not always be so. Dominant species alternate with less dominant species which are abundant after a disturbance and create a new stability in the communities. (Wohlgemuth et al. 2019: 39). Dominant species are less resistant to disturbance than other species (Wohlgemuth et al. 2019: 101). Therefore, disturbance triggers a re-organization of species. On the other hand, pathogens submit dominant plants and facilitate diversity (Wohlgemuth et al. 2019: 121). In the city the dominant species are urbanophilas, persistent, resistant to disturbance and they have a long vegetative period. In this sense, they are often the introduced trees.

Lima's predominant species include climax vegetation found in riparian ecosystems, such as Peruvian pepper and Humboldt's willow, which are now reoccupying the space (see Fig. 7.2). Desert saltgrass (*Distichlis spicata*) abounds in grassland communities in the the wetland ecosystem in the communities, while species of reed (*Thypha sp.*) prevail in cattail communities. In the parks of Lima, the predominant species of flora include trees such as eucalyptus, Peruvian pepper, and weeping figure.

7.5.2 Pioneer Species

These species occupy the space during the first face of succession facilitating space for the rest of the plants. They are plants with a short vegetative period and produce seeds fast and abundantly. Once the other species establish, the Pioneer plants are displaced since they have a low tolerance to competition between species. The role of Pioneer species is very important in the natural restauration of ecosystems. They are the first to arrive and install, leading to the arrival of new species. Many pioneer species are native and can occupy spaces in cement cracks, corners in pedestrian crossings, a crack near a carpark or a spot beneath a dripping pipe. We usually do not pay much attention to them and consider them undesired weed (see Fig. 7.4).

In beach condominiums or dry, abandoned parks in Lima's coastal districts, desert saltgrass and sea purslane (*Sesuvium portulacastrum*) are pioneer species; among the stones in humid areas one finds maidenhair fern (*Adiantum sp.*); red spinach (*Amarathus dubius*) grows along street edges; and the castor bean (*Ricinus communis*) thrives in dry, sandy areas by the riverside.

Fig. 7.4 A pioneer species dandelion (*Traxacum officinale*) growing between 2 stones. *Author* Ana Sabogal

7.5.3 Introduced Species

Introduced species are those that have not evolved in the same place they are planted coming from a different ecosystem. Urban flora is composed mainly by introduced species. There are many reasons for introducing a species in a city but the main one is ornamental. Most of the introduced species are potentially invasive. Since they are plague resistant, they often carry new pests or disease form their place of origin.

Introduced species carry pathogens which local species have no resistance for, so introduced species should develop new interspecific relationships with local species which also have pathogens. Since insects and pests are generalists, they adapt and distribute with more speed than plants, specially trees (Wohlgemuth et al. 2019: 198). Therefore, they are a serious threat for local species. However, in general there are multiple biotic and abiotic factors that determine the persistence of pests and disease and their installation as part of the trophic chain in the ecosystems (Wohlgemuth et al. 2019: 200). In the city, ecosystems are modified so plants, pests and even disease need to adapt.

The city can be refuge to many species (Endlicher et al. 2012: 163). There is a diversity of criteria to decide which plants to introduce. However, they should meet certain characteristics in terms of aesthetics, and they should be resistant to the abiotic stress characteristic of the city. Most parks and gardens in Lima have introduced species because Lima has very little rain. Species distribution also varies in the city. For example, in Berlin the city centre has 35% of introduced species whereas the periphery has 18% percent (Kowarik 1992 cit. Endlicher et al. 2012).

In the city, introduced species respond to ornamental criteria subject to fashion. In Lima, it is easy to note that every 10 years a new species of tree is given priority. Therefore, we can

Fig. 7.5 Sanssousi Garden, Potsdam. *Author* Ana Sabogal

determine when the park was made by the species and age of the oldest trees. In Lima, we can determine that the yellow trumpet bush (*Tecoma stans*) is distinctive of this decade, the California fan plam (*Washingtonia filifera*) and bottle tree (*Hyophorbe lagenicaulis*) are from the first decade of the twenty-first century, the ficus tree (*Ficus benjamina*) is distinctive of the 90 s, whereas the tipu tree (*Tipuana tipa*) is characteristic of the 80s.

Introduced plants are part of the development of parks and gardens. As in Lima, fashion for certain introduced plants is distinctive in the European parks and gardens. In this context, between the seventeenth and eighteenth century, greenhouses called "*orangerie*" were in fashion because citruses were brought and planted there from the middle east where the weather was warmer, so plants had to be protected in winter. This tendency was stronger in the nineteenth century with the development of the steel industry that enabled the construction of steel greenhouses (Kluckert 2000: 457). The most famous greenhouse is the Kew Garden designed by architects Decimus Buton and Richard Turner between 1845 and 1847 (Kluckert 2000: 457). Another important *Orangerie* is Sanssousi park in Potsdam in the outskirts of Berlin, Germany, where plants are transferred from the terraces to the *orangerie* during winter and fashionable palms were also introduced to the *orangerie*. Vides, characteristic of warmer climates, were planted in the terraces (see Fig. 7.5).

Cities are centres of importation and naturalization of species (Kowarik 2011 cit. Endlicher et al. 2012: 162) as well as evolution. (Endlicher et al. 2012: 162). Most species introduced to the city are adapted and resistant plants (Endlicher et al. 2012: 130). The tamarisks (*Tamarix sp.*) is a species introduced by the river with high distribution in the north of Peru reaching Lima whereas the eucalyptus (*Eucalyptus globulus*) grows easily with very little water in desertic areas. Meanwhile, a species present in almost all the parks of Lima is the weeping figure.

7.5.4 Native Species

Native species have developed in the same space they are from. In fact, all plants from the city are species introduced from another ecosystem since the ecosystem in the city is anthropic, different from the space where the native species are from. We should consider that the temperature in the city is 5 °C higher than in the countryside so for example, in Varsaw even though 20% of species are native, they are actually not since the climate in the city is different from the countryside where these species are from (Swoczyna et al. 2017). It is the same case for soil and air conditions.

In Lima city there are few native species. As explained in previous chapters, Lima was founded in the Rimac valley. However, great part of the city expanded into a desertic ecosystem so most of the species are introduced and their survival depends on watering so they need human intervention.

The native species of Lima include Amancay (*Ismene Amancaes*), which has been recultivated and is now found in some of the city's parks. Other species commonly encountered in parks are the aforementioned Peruvian pepper, and yellow elder (*Tecoma stans*).

7.5.5 Key Species

Key species are those that facilitate the conquest of space by other species. They offer shade, shelter, heat, nitrogen or other characteristics that enable the conquest of the space. Therefore, these species should be given priority when restoring the ecosystem. For example, legumes are key species; it is known that planting clover and alfalfa enrichens the soil with nitrogen and facilitates other species to develop. Unlike Pioneer species which are the first to install, key species can install at any stage and facilitate the development of the ecosystem. Key species are Urbanophilas with a short vegetative period.

Species that are key to Lima's park ecosystem are: Desert saltgrass (*Distichlis spicata*) and sea purslane (*Sesuvium portulacastrum*) in coastal areas, where the soil is sandy; various species of airplant (*Tillandsia sp.*) grow in wetlands on the xerophytic hills that surround Lima. Finally, wild cane (*Ginerium sagittatum*), and pájaro bobo (*Tessaria integrifolia*) grow along river banks (Brack and Mendiala 2000).

7.5.6 Indicator Species

Indicator species, as explained by its name, indicate the characteristic of their environment. These are especially important in the city since they show pollution problems that could be affecting plants and animals. One important group of pollution indicators are the lichens formed by algae symbiosis (see Fig. 7.6). These are especially sensitive to pollution. Likewise,

Fig. 7.6 Lichen as humidity indicator species. *Author* Ana Sabogal

the distinctive plants from the genus Tillandsia that hang from trees and cables also known as old man's beard. Seashore saltgrass (*Distichlis spicata*) or the shoreline purslane (*Sesuvium portulacastrum*) are indicators of salinity. Maidenhair (*Adiantum cuneatum*) is another indicator species that shows there is abundance of water. These species are especially susceptible to contamination, as will the reeds of the Tillandsia genus that hang from trees and cables in the more humid areas of Lima, giving rise to their local name "old man's beard." Indicators of soil salinity are desert saltgrass and sea purslane, encountered in beach ecosystems. Delta maidenhair (*Adiantum cuneatum*) is another indicator species, signalling the presence of abundant water in humid parts of Lima.

Indicator species are resistant to the peculiar characteristics of the environment. They are urbanoneutral with a short vegetative period.

7.5.7 Spontaneous Species

Spontaneous vegetation is the one that surges naturally, frequently in free spaces that are contaminated or degraded where no other species wants to be. They could grow in cracks between

Table 7.3 Classification of species: distinctive characteristics, own authorship

Classification of species		
Type of species	Distinctive characteristics	Example for Lima
Dominant species	Leads the development of the ecosystem Urbanophile, tolerant to disturbance, resistant, long vegetative period	Peruvian pepper (*Schinus molle*), Humboldt's willow (*Salix humboldtiana*), Desert saltgrass (*Distichlis spicata*)
Pioneer species	Little tolerance to competition, short vegetative period	Desert saltgrass (*Distichlis spicata*), sea purslane (*Sesuvium portulacastrum*), maidenhair fern (*Adiantum sp.*), red spinach (*Amarathus dubius*), castor bean (*Ricinus communis*)
Introduced species	Urbanophile, resistant, often dominant	Washingtonia palm (*Washingtonia filifera*), botle plam (*Hyophorbe lagenicaulis*), (*Ficus benjamina*), (*Tipuana tipa*)
Native species	Urbanoneutral o urbanopphoba Grows in its place of origin	Yellow elder (*Tecoma stans*), Peruvian pepper (*Schinus molle*), Humboldt's willow (*Salix humboldtiana*), amancay (*Ismene amancaes*), desert saltgrass (*Distichlis spicata*)
Key species	Urbanophile, short vegetative period	Desert ecosystem: desert saltgrass (*Distichlis spicata*), sea purslane (*Sesuvium portulacastrum*) Xerophitic ecosystem: airplant (*Tillandsia sp.*) Riverside: wild cane (*Ginerium sagittatum*), pájaro bobo (*Tessaria integrifolia*)
Indicator species	Urbanoneutral, resistant to the peculiar characteristics of the environment and with a short vegetative period	Airplant (*Tillandsia sp.*), saltgrass (*Distichlis spicata*), sea purslane (*Sesuvium portulacastrum*), maidenhair fern (*Adiantum cuneatum*)
Spontaneous species	Resistant, pioneer, short vegetative period	Yellow elder (*Tecoma stans*), sea purslane (*Sesuvium portulacastrum*), saltgrass (*Distichlis spicata*)

stones, areas polluted by underground water or soil degraded by construction. Spontaneous vegetation can vary depending on the type of soil and pH. They can also serve as indicators of the type of soil. In Lima, highly resistant species like the kikuyu grass (*Pennisetum clandestinum*), seashore saltgrass (*Distichlis spicata*) or shoreline purslane (*Sesuvium portulacastrum*) will grow in sandy soil. This type of vegetation can be seen in spaces near the sea or in areas like the "cedros de Villa" where there is great human and grazing pressure, whereas the Peruvian pepper tree (*Schinus molle*), elder (*Sambucus peruviana*) or the castor bean (*Ricinus communis*) will prefer the sandy loam. This type of vegetation can be found in the river forest along the river bank of the Rimac, Chillon or Lurin rivers since spontaneous vegetation grows at the edge of the rivers especially when surrounding population use it as a dump for fluids. There is a combination of native species (Peruvian pepper tree and willow) with introduced species like the castor-oil plant. In both cases, they are pioneer plants that constitute spontaneous vegetation in abandoned spaces. These plants will colonise the space and facilitate the way for the development of others without having been planted.

Native species can also germinate without being planted forming a spontaneous flora that reconquers the space, often facilitated by introduced species which can also form spontaneous species. This vegetation is formed by a distinctive group of plants that protect each other and grow together. Spontaneous species that grow in urban centres without being native are for example the dandelion (*Traxacum officinale*) or plantain (*Plantago major*) (Endlicher et al. 2012: 134). Native spontaneous species in Lima are for example: the yellow trumpet bush (*Tecoma stans*), the Peruvian pepper tree (*Schinus molle*) and the cattail (*Typha sp*). In the Table 7.3 the distinctive characteristics of the species are summarized.

References

Beatley T (2011) "Biophilic Cities" from Biophylic cities. In: Wheeler S, Beatley T (2014) The sustainable urban development, 3rd edn. Routledge, London and New York, pp 181–183

Calthorpe P (1993) "The Next American Metropolis from the next american metropolis: ecology, community, and the American Dream. In: Wheeler S, Beatley T (2014) The sustainable urban development, 3rd edn. Routledge, London and New York, pp 119–129

Endlicher W, Kraas F, Krajewski Ch (2012) Einführung in die Stadökologie. UTB, Stuttgart, p 272

Francis R, Chadwick M (2013) Urban ecosystems: Understanding the Human Environment. Routledge, USA, p 220

Hagan S (2015) Ecological Urbanism: the nature of the city. Routledge, Oxon, p 174

Hanna K, Slocombe S (2013) Sustainability and integrated approaches to regional planning. In: Dale A, Dushenko W, Robinson P (eds) (2013) Urban sustainability: reconnecting space. University of Toronto Press. Toronto, Buffalo, Londo, p 286 P: 27–49

Kluckert E (2000) Grandes jardines de Europa: desde la antigüedad hasta nuestros días. Könneman, Köln, p 496

Kowarik I (2003) Biologische Invationen: Neophyten und Neozoen in Mitteleuropa. Ulmer, p 380

Steiner F (2011) Landscape Ecological Urbanism from landscape and urban planning. In: Wheeler S, Beatley T (eds) (2014) The sustainable urban development, 3rd edn. Routledge, London and New York, pp 190–194

Swoczyna T, Borowski J, Latoja P (2017) Trees and shrubs for urban plantings: introduced or native species. In: Congres presentation: problems of landscape protection and management in XXI century. Organized by Warsaw University of Life Sciences, Polski klub ekolgiszny

Turner M (2001) Landscape ecology: in theory a practice pattern and process. Springer, New York, p 401p

Wohlgemuth T, Jentsch A, Seidl R (eds) (2019) Störungsökologie. Utb Haupt Verlag, Gernany, p 396

8 Ecological Restoration

Abstract

This chapter defines ecological restoration distinguishing it from ecosystem re-naturalisation, based on the fact that urban ecosystems are new and different from the original. It describes the characteristics of urban ecosystems and species focusing the description on Lima city, linking the description to the origin of species and pointing out that great part of the fauna species in Lima comes from different origins. It examines urban ecosystem resilience and adaptation to new species as well as the re-conquest of abandoned spaces and those destroyed by disturbance. It analyses implications of the disturbance on ecosystem dynamics in the light of an urban ecosystem in Lima city. It discusses ecosystem resilience and factors that facilitate it considering that the results of the disturbance can originate changes in the ecosystem and in species dynamics. It explains how ecosystems are expected to change in Lima in view of climate change. Restoration concepts are applied and analysed together with its implications on species and ecosystems in Lima city. This chapter centred its exposition on the ecological restoration to ensure the quality of the green space and the ecosystem's resilience. In this way, it linked the Sustainable Development Goal 11 and 15 related to reverse lad degradation.

Keywords

Ecological restoration · Disturbance · Resilience · Climate change · Lima

This chapter defines ecological restoration distinguishing it from ecosystem re-naturalisation, based on the fact that urban ecosystems are new and different from the original. It describes the characteristics of urban ecosystems and species focusing the description on Lima city, linking the description to the origin of species and pointing out that great part of the fauna species in Lima comes from different origins. It examines urban ecosystem resilience and adaptation to new species as well as the re-conquest of abandoned spaces and those destroyed by disturbance. It analyses implications of the disturbance on ecosystem dynamics in the light of an urban ecosystem in Lima city. It discusses ecosystem resilience and factors that facilitate it considering that the results of the disturbance can originate changes in the ecosystem and in species dynamics. It explains how ecosystems are expected to change in Lima in view of climate change. Restoration concepts are applied and analysed together with its implications on species and ecosystems in Lima city. This chapter centred its exposition on the ecological restoration to ensure the quality of the green space and the ecosystem's resilience. In this way, it linked the Sustainable Development Goal 11 and 15 related to reverse lad degradation.

A. Sabogal, *Urban Ecology*, Sustainable Development Goals Series,
https://doi.org/10.1007/978-3-030-69905-5_8

8.1 Ecological Restoration

Ecological restoration is the scientific study of the processes to assist in the recovery of degraded ecosystems which goes in hand with the re-naturalisation of the ecosystems. The re-naturalisation of ecosystems contemplates the restoration of ecosystems by also considering social intervention in the process to facilitate changes (Zerbe 2019: 4). In this sense, ecological restoration proposes the return to natural ecosystems. In the world, there are 2000 million hectares of deforested and degraded landscape which should be recovered (CEPAL et al. 2017: 144). Part of them are the urban ecosystems, which are reconstructed ecosystems composed by species from different parts of the world which were not together in their original ecosystem. In the current context when cities more than ever have the greatest impact both in terms of world climate and population distribution, restoration and re-naturalisation of urban ecosystems constitute a priority. The recovery processes could contribute to reduce poverty, improve food security, mitigate the effects of climate change, conserve biodiversity, improve soil and water protection, and increase forest surface (CEPAL et al. 2017: 144). Urban forestry can contribute to this. Then, cities form new ecosystems to replace a degraded one by another in which human beings have the primary role. For this reason, urban ecosystems in their essence are degraded ecosystems where identical restoration is not possible. It is necessary to think about how to restore the ecosystem by integrating it to the space or by forming a completely new ecosystem different from the original one. The distinctive Characteristics of the Anthropic and natural ecosystem are in the Table 8.1. Ecosystem re-naturalisation and restoration has to consider the population's interests and participation so that urban ecosystems can be pleasant and useful for the citizens.

Urban ecosystems constitute new organisations and ecological systems in which both processes and species form unknown dynamical systems. They develop new interspecific relationships, new trophic chains and new ecological niches. In this sense, to restore urban ecosystems it is important to consider that these ecosystems are ignored and novel, so native species will not necessarily develop; They will need to adapt to the new conditions of physical factors and compete with new species. In this case, both the substrates used, and the introduced plants constitute new ecosystems (Kowarik 2011 cit. Zerbe 2019: 433). Soil and plants are very different from the original ones. Soil is often compact, with poor water flow, higher levels of alkalinity, more nutrients, or they could also be contaminated (Rebel 2009 cit. Zerbe 2019: 422). Urban ecosystems often have more biodiversity (Sukopp and Wittig 1998, Berninde et al. 2015 Zerbe et al. 2003 all cit. Zerbe 2019: 422). Furthermore, the diversity of spaces causes a larger diversity of ecosystems in reduced spaces (see Table 8.2 and Fig. 8.1). Each one of these spaces has different abiotic conditions which can contain a diversity of flora and fauna forming mosaics with a variety of ecosystems.

To achieve sustainable development in Lima and spatial justice in its public spaces and parks,

Table 8.1 Characterisation of urban ecosystems, own authorship

Ecosystem characterisation	
Ecosystem	Distinctive characteristics
Natural	Native species, local soil, local fauna, medium biodiversity
Urban	Introduced urbanophile species, reconstituted azonal soil, fauna formed basically by generalists combining native and renaturalised introduced species, high biodiversity
Restored	Native species, zonal soil, local fauna, medium biodiversity
Renaturalised	Native and introduced species, azonal soil, local and introduced fauna, high biodiversity

Table 8.2 Characteristic ecosystems in Lima city, by the author

Characteristic ecosystems in Lima city	
Ecosystem	Peculiar characteristic that defines the space
Lima parks	Reconstituted soil, combination of wild and introduced species
Private gardens	Large number of introduced species
Industrial zones	Contaminated soil, introduced species
Densely constructed zones	Soil compaction
Residential Zones	Reconstituted soil
Cemeteries	Abundancy of water
High transit pedestrian zones	Soil Compaction
Avenues	Air pollution
Wildlife Sanctuary Villa Marshes	Wetland, native species
Costal	Salt in the air and soil
Roof tops	Sun incidence and low humidity
Bio-orchards	Reconstructed soil and exotic vegetables

Fig. 8.1 White tail deer (*Odocoileus virginianus*). *Author* Ana Sabogal

it is important to consider urban ecology, given that these are new ecosystems that must be restored in line with regulations. Lima's urban ecosystem includes the following distinctive soil characteristics: sandy soil in the areas close to the sea; compact, acidic soil due to construction in densely populated and old parts of the city, such as the center; soil contaminated by industry in the east; and marshy soil in the south. Table 8.2 shows the diversity of Lima's ecosystems.

8.2 Urban Species

Most of the species in the city are introduced by the population. Practically all vegetation is planted, whereas in the case of fauna, native species mix with domesticated ones (see Table 7.3 and Fig. 7.1). Spontaneous species in the city are small and resistant ruderal species. Many of them are urbanophiles since they prefer the

Table 8.3 Classification of the fauna in Lima based on origin, own authorship

Classification of the fauna in Lima based on origin	
Species	Origin
Pacific parrotlet (*Forpus coelesteris*)	Domestic, dry forest in the north of Peru
Guayaquil squirrel (*Sciurus stramineus*)	Migration from the dry forest in the north of Peru
Saffron Finch (*Sicalis flaveola*)	Domestic, from the Amazonian basin
Blue-Grey Tanager (*Thraupis episcopus*)	Domestic, from the Amazonian basin
White tail deer (*Odocoileus virginianus*)	Domestic, from the dry forest in the north of Peru
Budgie (*Melopsittacus undulatus*)	Domestic, from Australia
Blue-black grassquit (*Volatinia jacarina*)	Habitat in the low Andean slope
Groove- billed ani (*Crotophaga sulcirostris*)	Original, of ample distribution
Peregrine Falcon (*Falco peregrinus*)	Original, adapted to the city
Amazilia hummingbird (*Amazilia amazilia*)	Original, adapted to the city
Burrowing owl (*Athene cunicularia*)	Original, only found in clearings
Peruvian thick knee (*Burhinus superciliaris*)	Original, only found in clearings
Blue-and-white swallow (*Pygochelidon cyanoleuca peuviana*)	Original from basins, adapted to the city
Peru coast toad (*Rhinella limensis*)	Original from basins, almost disappeared from Lima due to pollution
Lima leaf-toed gecko (*Phyllodactylus sentosus*)	Original, in critical danger

(*Source of data* SERFOR 2019)

environment in the city (Wittig et al. 1985 cit. Zerbe 2019: 423). Some of the species can only be found in anthropic spaces (Zerbe 2019: 423).

In case of the fauna in Lima city, we note that it is formed by released domestic animals adapted to the parks of Lima where they have become wild again. Such is the case of the Pacific parrotlet (*Forpus coelesteris*), native from the dry forest in the north of Peru. On the other hand, other species have come a long way from the north of Peru during the phenomena of "El Niño", finding in Lima a suitable habitat. Such is the case of the Guayaquil squirrel (*Sciurus stramineus*), also native from the dry forest in the north of Peru (see Sect. 6.1). These species can be used to restore ecosystems in Lima and other cities along the coast of Peru with similar ecosystems. Many native species are displaced by introduced species and climate change distinctive of the city. Such is the case of the Lima leaf-toed gecko (*Phyllodactylus sentosus*) or the Peru coast toad (*Rhinella limensis*). But also, because they are sensitive species afraid of the population like the bird called Peruvian thick knee (*Burhinus superciliaris*) which flies away whenever approached by people, or the burrowing owl (*Athene cunicularia*), which is found with more difficulty since it nests in the sand in clearings.

The Table 8.3 also shows Lima's most representative fauna by origin, most being introduced species.

Urban ecosystems typically have a lot of biodiversity. As we can see in Table 8.3, there are species from the original ecosystem as well as new species from a diversity of ecosystems. In this type of ecosystem, the formation process accelerates and stabilises more quickly than natural ecosystems. To restore ecosystem processes it is necessary to consider that pioneer species will set the foundations for the new ecosystem and will facilitate the development of the trophic chain in the city. Only when the trophic chain is formed, will the urban ecosystem stabilise.

The resilience of natural ecosystems depends on the upkeep of the city's remaining natural spaces, where endemic species can find sanctuary; examples include the Lima toad (*Rhinella limensiS*), the Lima leaf-toad gecko (*Phyllodactylus sentosus*) and the burrowing owl (*Athene cunicularia*) (see Table 8.2). The resilience of the natural ecosystems will depend on the remnants of the natural spaces where native species like the Peru coast toad, the Lima leaf-toed gecko or the burrowing owl can find refuge (see Table 8.3). This is called functional resilience. In the cities the functional resilience of the ecosystems increases (Wohlgemuth et al. 2019: 314). Spaces, often overlooked, like the sand expanses outside the city or abandoned spaces where nature reconquers its living conditions can be vital to attain restoration; species that persist in them will become pioneers in the reconquest of the ecosystem. However, there are constant disturbances and variations in biodiversity. To face this, urban ecosystems will have variable and unpredictable responses not always giving priority to native species but to those that can adapt better and reconquer the space faster facilitating the process of ecosystem restoration. Nevertheless, when disturbances are excessive, as often happens in the city, ecosystems do not stabilise, and they are constantly reinventing themselves to face the permanent introduction of new species and changing abiotic factors. In this process, the presence of humans is essential since they introduce the factors necessary for species survival such as soil, water and plants, facilitating the process for some species more than others and, therefore, determining to a great extent species selection in the ecosystem.

8.3 Ecosystem Disturbance

Ecosystem disturbance is defined as changes in space and time that lead to the decrease in biomass and availability of resources in living communities; the more intense the disturbance, the more repercussions it will have on the biomass (Wohlgemuth et al. 2019: 24). These changes are discreet. However, they produce great alterations, yet, ecosystems are accustomed to these alterations after which they can reorganize. The capacity to reorganize is known as ecosystem resilience (Wohlgemuth et al. 2019: 91). The more accustomed the ecosystem is to alterations caused by disturbances, the more its capacity of resilience which is also improved when alterations are similar. In this sense, Rimac river is accustomed to the annual changes of current and the ecosystems in the north of Peru are accustomed to “El Niño” phenomena. Urban ecosystems are constantly submitted to anthropic disturbance.

Urban ecosystems experience constant stress due to the constant changes to which they are exposed. Indeed, constant changes over time that do not affect ecosystem structures are not disturbances but stress (Wohlgemuth et al. 2019: 24). Thus, for example, intensive pruning or irrigation cause stress that can destroy some plants and cause others to rot. Conversely, examples of disturbance might include the Rímac busting its banks, soil compacting for construction, and spillage of cooking oil into urban soil by restaurants. These are changes to the ecosystem structure. Disturbances may be cyclical, and part of the ecosystem's characteristics; or cascading, causing cascade reactions down the ecosystem that change over time and according to the form of the disturbance (Wohlgemuth et al. 2019: 29). They are therefore foreseeable, and the ecosystem may have difficulty restructuring. Cascade reactions are more common in urban ecosystems where soil conditions are highly changeable. Thus, the ecological succession of the city is unpredictable and non-linear given the large numbers of disruptive factors (Francis and Chadwick 2013: 118). Cascade reactions are non-cyclical, constantly altering the structure of the ecosystem. Ecosystems that survive are those made up of resistant plants, most of which are introduced species.

Urban ecosystems are constantly submitted to stress caused by constant changes. Continuous changes that do not have repercussions on the ecosystem’s structure are not considered disturbances but stress (Wohlgemuth et al. 2019: 24). In this sense, stress for example, is caused by pruning or intense watering that affects plants and could cause rotting, but a disturbance is the flooding of river Rimac or soil compaction for construction or an oil spill in urban soil caused by the presence of nearby restaurants. Disturbances can be cyclic, and they can be part of the ecosystem's characteristics or they can cascade and produce a chain effect in the ecosystem. Cascade reactions change in time and the disturbance changes form (Wohlgemuth et al. 2019: 29). Therefore, they are less foreseeable, so the ecosystem has more problems to restructure. They are more common in urban ecosystems where soil conditions are highly changeable. Due to the great variety of disturbances, ecological succession is vital and non-linear (Francis and Chadwick 2013: 118). Disturbances that cascade do not occur cyclically and constantly alter the ecosystem's structure. The surviving ecosystems are formed by resistant plants, most of them introduced.

Disturbances balances ecosystems and enable their renewal. An unaltered ecosystem could collapse due to species dominance and lack of competition. The Suppression Hypothesis affirms that disturbances will be more intense, the fewer the frequency of the disturbances. Therefore, the intensity and frequency of disturbances are inversely correlated. When disturbances are strong, the ecosystem requires more time to recover (Wohlgemuth et al. 2019: 32). A disturbance may imply changes in the availability of resources as well as in the dominance of species (see Sect. 6.2). After the disturbance, species need to readjust; their roles may change as well as the ecosystem's dynamics. The species that survive disturbances are usually not resistant to competition with other species, but they are resistant and adaptable to these disturbances (Wohlgemuth et al. 2019: 33). These species are resistant to environmental changes that constantly occur in the city, which as we mentioned above, most cases are introduced species.

Urban ecosystems are essentially young since they are constantly being disturbed and consequently renewed. However, these are different from the natural ecosystems in the sense that the species in each stage of development do not correspond to the classical ecological succession that typically have pioneer species of a short vegetative period in the initial stage, followed by seral stages and trees during climax (see Sect. 7.3 and Fig. 8.2). In an urban ecosystem, a tree can mark the beginning of a park ecosystem. Even though stages can be skipped, and species do not have to be typical of each stage, the ecosystem must go through processes and adjustments to reach a balance distinctive of the climax stage. For this reason, the pioneer species in this type of ecosystems are not usually the same as the ones in natural ecosystems. The fauna and flora require neither species development nor soil enrichment. In residential areas soil will have more nitrogen (Zerbe 2019: 425). This enables the development of new ecosystems, with vegetation composed by ruderal pioneer species of fast, underground (threofita) growth (Zerbe 2019: 426). In Lima, such is the case of areas at the

Fig. 8.2 Urban Ecological succession. *Author* Ana Sabogal, Designer: Juan Pablo Bruno

Table 8.4 Effects of the disturbance in ecosystem dynamics, own authorship

Effects of the disturbance in ecosystem dynamics				
Factor	Degree	Frequency	Alteration	Ecosystem
Disturbance	Low degree	High frequency	Biotic factor	Ecosystem resilience and restoration
Disturbance	High degree	Low frequency	Abiotic factor	New ecosystem
Disturbance	High degree	High frequency	Biotic and abiotic factor	Pioneer ecosystem

edge of the streets or avenues where the amaranth grows (*Amarathus dubius*) or in dry sandy areas the castor oil plant or the seashore saltgrass which can be found in condominiums and districts near the ocean. All these ruderal plants are resistant to the new conditions in the city and mark the beginning of the ecosystem facilitating the process of formation. However, even though they can represent the pioneer stage, they can be replaced in parks and gardens by introduced species planted by humans, triggering the seral stages or a secondary succession.

The enrichment of city soil with nitrates and phosphates is caused mainly by the sewages that supply the soil and the groundwater with nitrogen and phosphorus. However, pollution and contamination of groundwater carries heavy metals and other pollutants with this nitrogen and phosphorus (see Chap. 6). Under these conditions, especially in industrial zones with factories or warehouses like in great part of Callao, plants must be resistant and capable to adapt to pollution, but few species can survive. However, some species like the castor oil plant (*Ricinus communis*), or the amaranth (*Amarathus dubius*) could survive, but, if we consider ornamental plants, we could use resistant plants like the Palo Verde (*Parkinsonia aculeata*), the Humboldt willow (*Salix humboldtiana),* or the pepper tree (*Schinus molle*) among a few others.

Disturbance regimes can be altered by human intervention by means of ecological engineering. However, this is extremely dangerous because it could intensify and accumulate the effects of the disturbance. For example, when we prevent fire in a forest ecosystem where it is part of the natural process, we could intensify the disturbance in the long run (Wohlgemuth et al. 2019: 32). In case of the city, if construction or engineering does not foresee disturbances or acts without considering them, it is not only extremely risky, but the risk is accumulative. Therefore, we must consider disturbances as part of the dynamic of space (see Table 8.4). Furthermore, even though environmental engineering can solve disturbance problems, the forces triggered by them are highly risky. In Lima, such is the case of the constructions of dykes without vents into the rivers like river Rimac with a variable interannual river course or construction of house without considering the regulated distance from the sea.

Landscape is changing and readjusting constantly. That is why we can talk about the dynamic balance of ecosystems which implies constant changes and adjustments of both the landscape and the composition of the species that form it. This implies that the Biological Structure of the ecosystem also changes and adjusts constantly (see Sect. 7.2). Dynamic equilibrium enables the subsistence of species showing stable and variable phases (Wohlgemuth et al. 2019: 36). In order to restore ecosystems, it is essential to understand and include these processes in the restoration. Landscape restoration depends on two aspects of the disturbance: dimension in relation to the ecosystem's size and time both in terms of duration and frequency. Additionally, fragmentation also has an effect. If the ecosystem is fragmented, the disturbance will have spatial limitations (Wohlgemuth et al. 2019: 36). On one hand, isolated spaces do not allow species to migrate. On the other hand, they might result in the survival of some populations that would be affected by interconnected ecosystems. Small isolated spaces can be a sanctuary and harbor niches. In these spaces, there is less interspecific competition and there are more abiotic factors available for species to survive when facing disturbance (Reif and Achtziger 2000 cit. Wohlgemuth et al. 2019: 293). In the city, most of the spaces are fragmented with little or no connection, especially in Lima where the ecological corridors are few (see Sect. 6.2). Therefore, the disturbance of a space, with a few exceptions, will not affect other ecosystems. This has advantages and disadvantages in terms of refuge since some species will not find sanctuary in case of disturbance. Such is the case of birds of Wildlife Refuge in Lima or species of native plants after the fertilization of parks. In this context, and especially in the context of the desertic ecosystem in most of Lima, human intervention is key to facilitate the restoration of ecosystems.

The possibility and degree of restoration of an ecosystem depends on the degree of deterioration and modification which is directly related to the age of the city. The process of urbanization is accelerated making ecosystem adaptation and conservation difficult. Species will react differently when facing stress modifying the physical and biological structure of the communities and setting course to a new succession. In old cities, ecosystems are modified and transformed in a state of succession of its own, different from natural succession; in this process, species have incorporated native and introduced species. Restoration implies having the knowledge about the initial dynamic of nature as well as the disturbance process.

Many cities have restored and reconquered their natural spaces. Such is the case of the natural channels in San Antonio, Texas (Riley 1998: 189) as well as the recovery of Seoul river source, where ecological corridors were recovered by uncovering the Cheonggyecheon river, overcrowded by the city (see Fig. 8.3). By doing so, it was possible to develop a new space, an ecological corridor, that integrates the river to its design (see Sect. 4.4). Another example is the Kienberg park in Berlin, also known as the Park of the Future. In times of the Democratic Republic of Germany an old channel was covered, but, time after it was restored. Currently, after 30 years, natural wetlands have regenerated by means of secondary succession, enabling native species to reconquer their space (see Fig. 8.4).

Fig. 8.3 Cheonggyecheon river, Seoul, Corea. *Author* Ana Sabogal

Fig. 8.4 Wetlands in Kienberg or Back to the Future park, Berlin. *Author* Ana Sabogal

Most of the time, natural disturbance does not occur uniformly, creating a heterogeneous landscape with spots (Turner 2010 cit. Wohlgemuth et al. 2019: 278). The same occurs with anthropic disturbance; when it is abiotic and disturbs soil and air, it is harder for the ecosystem to recover. On the other hand, when the disturbance is biotic, the ecosystem will recover in no time unless the disturbance is significant. In this sense, during pruning trees will have to adjust to new conditions, since the plants beneath the trees will get more light. This will also occur if a grove is pruned, which will affect the size of each tree crown as well as the competition between trees and will have consequences in the number of nests each tree will hold, among many other alterations. Even though positive disturbances prevail, abiotic alterations can trigger biotic effects (Siedl et al. 2017 cit. Wohlgemuth et al. 2019: 332). This chain effects are often difficult to predict and estimate. For example, alterations to the landscape caused by giving a soil a different use or building, alter the pH affecting both the fauna in the soil as well as the plants. Pruning of an old tree can also produce a chain reaction since it has repercussions on the fauna. Old trees affect microhabitats which are often inhabited by small animals (Vuidot et al. 2011, Larrieu and Cabanettes, 2112 both cit. Wohlgemuth et al. 2019: 289) (see Fig. 8.5). Among animals living in trees in Lima we can find birds, mice, squirrels and many more, so, obviously, tree pruning, elimination or any form of intervention will affect the fauna inhabiting the tree.

8.4 Resilience

Ecological resilience is the capacity of an ecosystem to recover naturally in response to a disturbance. Resilience is natural when the ecosystem reacts to changes and reorganizes. However, ecosystem resilience can be promoted. In this sense, there are three types of resilience: technical resilience, ecological resilience and socio-ecological resilience (Wohlgemuth et al. 2019: 93). Technical resilience is focused on restoring the factors that define the ecosystem's balance centered on technical aspects and treating each factor separately. Technical resilience is based on human intervention to restore the ecosystem's balance. Technical resilience defines restoration as a linear process in which factors are recovered one by one until the previous ecosystem is recovered. On the other hand, ecological resilience studies and involves

Fig. 8.5 Cavity in the destroyed stair, habitat to a diversity of fauna species. *Author* Ana Sabogal

changes in the composition of the ecosystem produced as a result of the disturbance and restores the functions of the ecosystem. It considers that ecosystems do not necessarily need to return to the same succession. Ecological resilience is the response of the ecosystem to changes and disturbances that enables the ecosystem to return to a state of balance which is not necessarily the same one there was before the disturbance. When the disturbance exceeds the ecosystem's ecological resilience, its processes and development are modified (Wohlgemuth et al. 2019: 93). Then, a new ecosystem is formed with different characteristics from the previous one. Finally, socio-ecological resilience is the social and political capacity to respond to the changes in the ecosystem (Adger 2000 cit. Wohlgemuth et al. 2019: 94). This implies the understanding of the changes and the incorporation of politics and use of space in order to balance the ecosystem.

The capacity of an ecosystem to regenerate and reinvent itself will depend on many factors. This combination of factors has different stages. The ecosystem balance depends basically on the interaction of factors that conform it. Restoration can establish new interactions even with the same factors (Holling and Gunderson 2002 cit. Wohlgemuth et al. 2019: 95). The regeneration and restoration of an altered ecosystem is much faster than the natural development of the ecosystem (Wohlgemuth et al. 2019: 95). Plants adapt to the alteration factors and will regenerate more easily if they are repetitive (Wohlgemuth et al. 2019: 100). However, there are ecosystems that take longer to return to its previous condition. To be able to determine how resilient an ecosystem is, it is essential to know its history and state of balance, defined specifically by the historical range of variability which enables us to predict the possible time and frequency of the new event that will alter the ecosystem (Keane et al. 2009 cit. Wohlgemuthet al. 2019: 102). Ecosystems which have not suffered a disturbance during their development will be less resilient and in case of a significant disturbance, the ecosystem could collapse without the intervention of technical resilience.

Ecosystem resilience depends on the degree of the disturbance. If it alters the abiotic factors, restoration will be more difficult, and in many cases, it will depend on technical resilience. Abiotic factors include water, air and soil. For example, groundwater is greatly altered by roads and drainage systems (Wohlgemuth et al. 2019: 291). Vegetation can be greatly affected which

Fig. 8.6 Spontaneous growth of poplar trees (*Populus sp.*) im Naturalpark Südgelände, Berlin. *Author* Ana Sabogal

can be very damaging in case of Sanctuaries of Wildlife like the Villa Marshes.

The spatial extension of the disturbance will define the possibility and time required by the ecosystem to recover. Thus, there is a spatial relationship which means that the bigger the expansion of the disturbance, and consequently the space of repercussion, less will be the resilience and slower will be the recovery (Dai et al. 2013, cit. Wohlgemuth et al. 2019: 104). Changes can often be detected first in a very small scale and then they can develop in a larger space. In this way, the increase of spatial diversity can foretell a change in the ecosystem (Kéfi et al. 2014 cit. Wohlgemuth et al. 2019: 104), which implies that species need to redefine their roles. It is essential to intervene at this point before the functional processes of the ecosystem are altered. In the city, spatial changes occur constantly, and ecosystems will have to reinvent themselves or modify completely with anthropic assistance. However, the constant destruction of the ecosystems makes the city's ecosystem resilience process more difficult so there is more dependency on external factors such as manure or chemical fertilisers which makes ecosystem balance harder to achieve. However, it will be able to develop with constant external support, despite its lack of resilience.

The natural park Südgelände in Berlin is an interesting park were spontaneous vegetation of poplar (*Populus sp.*) found the possibility to return and reconquest the area through the park protection. The new ecosystem builds a secondary succession in there the poplar trees are the key species to reach the ecosystem restauration trough the secondary succession (see Fig. 8.6).

This is the case of most of Lima's parks, where agrochemicals, excessive pruning, and poor management are ever-presents that contribute to disturbances. As a result, many municipalities have to replant trees and add agrochemicals anew. Despite rather than because of these actions, ecosystems become more resilient, preserving the species most resistant to change while municipalities constantly introduce new plants.

8.5 Effects of Disturbance in Species Dynamic

The Intermediate Disturbance Hypothesis suggests the need of a disturbance in a moderate intensity so an ecosystem can continue functioning. This hypothesis considers two premises. On one hand, the existence of a hierarchy of species and on the other hand, the tiered development of species adapting to the trade-off of roles during competition and tolerance to disturbance. In this way, we can identify species tolerant to disturbance which can prosper with limited resources (Wohlgemuth et al. 2019: 40). Pioneer species that occupy space in the first stage after the disturbance are not very tolerant to competition. In the second stage, they will be replaced by dominant species tolerant to competition between species (Wohlgemuth et al. 2019: 41). However, these pioneer species will be tolerant to the disturbance and will enable the restoration of the new ecosystem after the disturbance.

The disturbance also modifies spatial heterogeneity which has an influence on the diversity of ecological niches, and correspondingly has repercussions on the increase of species diversity (Wohlgemuth et al. 2019: 76). The disturbance does not only cause changes in species dynamic but there is also an effect of species dynamic on the disturbance (Hughes et al. 2007 cit. Wohlgemuth et al. 2019: 76). In the natural ecosystems of Lima, filter plant species such as achira (*Canna edulis*) and cattail (*Typha sp.*) protect water bodies. Others, such as desert saltgrass (*Distichlis spicata*) and sea purslane (*Sesuvium portulacastrum*) form dunes and prevent sand drift, thus protecting beach properties. Species like the Achira (*Canna edulis*) or the cattail (*Typha sp.*) will protect the water of ecosystems since they are filtering plants. Others like the seashore salt grass (*Distichlis spicata*) or the shoreline purslane (*Sesuvium portulacastrum*) will form dunes and prevent beaches from blowing away. Therefore, they protect houses from the sand. Little diversity is a consequence of high productivity and low disturbance (Huston 1994, cit. Wohlgemuth et al. 2019: 78). This is frequent in emerging ecosystems where diversity is still not high like the dune ecosystems formed by shoreline purslane (*Sesuvium portulacastrum*) as a dominant species accompanied by very few other species. In ecosystems subject to few disturbances, climax species dominate. These are dominant species with a long vegetative period and capable of competition with other species (Wohlgemuth et al. 2019: 78). Whereas elevated ecosystem productivity promotes dominant species which have won the interspecific competition, a high level of disturbances, both in terms of degree and frequency, promotes pioneer species and low diversity (Wohlgemuth et al. 2019: 78). In the climax stage, dominated by trees over other types of plants the ecosystem is in balance. If trees are felled or eliminated, the ecosystem will have to reconstitute from a starting point with pioneer plants.

Alteration modifies species distribution by changing the distribution of resources. When abiotic conditions are modified, they influence biotic conditions such as the availability of nitrogen, water and light, enabling or facilitating the development of new species (Wohlgemuth et al. 2019: 80). In the city, alterations of biotic conditions are often significant, and they can have repercussions on the ecosystems balance and the species dynamic. In cities like Lima almost all species are introduced, and native species will have to modify their living conditions in order to survive. Therefore, we only find resilient native plants or some more sensitive native plants which have been planted.

Disturbances modify the temporal succession processes, eliminating inertia of vegetation communities and facilitating change (Wohlgemuth et al. 2019: 82). They modify the vegetation communities facilitating the distribution of foreign species (Zonneveld 1995 cit. Wohlgemuth et al. 2019: 83). Then, new interspecific relationships develop. This dynamic is facilitated by ecosystem management such as parks where biotic and abiotic conditions are constantly modified.

Disturbances force natural selection that leads to the distinction of ecological niches in the communities modifying the species spectrum. What's more, the change in the availability of resources affects dominance which means that dominated species might become dominant so domination will be restructured in the community (Wohlgemuth et al. 2019: 39). There is a direct relationship between the diversity of alterations and the diversity of species. The greater the diversity in the alteration regime, the greater the diversity of species in the ecosystem (Wohlgemuth et al. 2019: 81). Furthermore, landscapes with the most disturbances can host more stable communities (Wohlgemuth et al. 2019: 81) that are used to these disturbances.

Species survival and their capacity to adapt depends on the availability of nutrients, growth speed and the capacity of plants to regenerate (Wohlgemuth et al. 2019: 124). Plant adaptation to face disturbance can be of two types: survival strategies or production of descendants (Wohlgemuth et al. 2019: 122). Therefore, plants either produce organs to ensure their survival for example, by producing more stolon's to migrate in search of food, they roll up and reduce their number of leaves in order to reduce evaporation or they use all their nutrients to produce seeds and die, leaving a new generation that will germinate when the environmental conditions are adequate. Often city ecosystems and plants need to face adverse factors which will make resilience strategies necessary.

Disturbances are then natural in urban ecosystems and species will adapt to constant changes. This type of ecosystems promotes the development of resistant and resilient species that displace the weaker ones. In order to win this competition, species must be of fast growth and development, resistant to city pollution and resistant to constant changes of abiotic factors. In many cases, species must be resistant to domestic animals, especially cats and dogs, pedestrians and children's playing. These are dominant species that could modify the roles of species and determine the survival of the other species in the ecosystem. Thus, resistent species such as weeping fig (*Ficus benjamina*) tend to be planted on streets; eucalyptus (*Eucalyptus globulus*) in parks; and mousehole trees (*Mioporum laetum*) close to the sea.

8.6 Climate Change

All ecosystems are currently affected by climate change. The ecosystems' capacity to respond to it is directly related to its capacity to respond to disturbances and consequently to their resilience. On the other hand, each species has ranges of factors that will enable their development. When the factors escape these ranges, the species suffer the consequences. The ranges applicable to each species are called ecological niches. Responses to Climate Change are not linear. However, they respond to the range of tolerance of the plants (Wohlgemuth et al. 2019: 325). In subtropical zones, changes in hydric regimes are expected (Wohlgemuth et al. 2019: 329). It is estimated that the temperature will increase between 1 °C and 3,7 °C (IPCC 2013 cit. Wohlgemuth et al. 2019: 326). Such is the case of Peru, specifically in Lima, initially there will be an increase in the water because the ice in the basins is melting which will lead to a new scenario.

As foreseen, Climate Change has caused an increase in the temperature in Lima and, therefore, an increase in rainfall for urban ecosystems. Locally Climate Change could be seen positively. The rise in temperature has a direct effect on the plants' speed of growth and water absorption. However, a higher temperature may intensify the number of pests and more humidity may trigger disease. This may cause an increase in the use of chemical products that pollutes groundwater. On the other hand, since the course of river basins is highly variable with tendency to flood during flood season, consequences could be catastrophic if risks are not prevented. All of this has direct consequences on the health of the urban population which should be foreseen in a timely manner. In Table 8.5, we can see the risks of Climate Change for Lima city.

Table 8.5 Disturbances and risks of climate change for Lima city, own authorship

Disturbances and risks of climate change for Lima city	
Disturbance	Risk
Change of river course	River flood. Population affected
Increase of pests and disease	Larger amounts of fertilisers and insecticides in parks, damage to health
Increase of rainfall	Increase of spontaneous vegetation, ecosystem improvement
Increase of rainfall	Increase of acid rain and damage to health
Increase of groundwater	Improvement of natural ecosystems

8.7 Practical Application in Lima City

Disturbance dynamics is directly related with the diversity of species and spaces. Rivers are typically dynamic spaces subject to constant changes, disturbances and subsequent regeneration of ecosystems. In these azonal spaces, soil is constantly carried by the river current and replaced by azonal soil from the basin higher up. However, in a river ecosystem there are communities protected from the current that can develop a little more lasting and stable dynamic which could lead to a zonal development. These are communities adapted to constant disturbances. Spaces like riverbanks, like those of the Rimac river which are pioneer spaces in terms of ecological succession since they are subject to constant changes and modifications since they are exposed to the variable course of the river (Fig. 8.7). Here, very resistant species have developed and adapted to disturbances. These species could be used for the re-naturalisation of ecosystems.

Among the factors that define river ecosystems one of the main ones is the speed of the current. Together with it, there are small spaces that form dynamic communities (Wohlgemuth et al. 2019: 37). Along its course, the river transports pieces of roots or branches that could germinate. Zones with fast current alternate with sedimentary zones, depending on the current, the geology, the landscape, the landscape and the

Fig. 8.7 Rimac river. *Author* Ana Sabogal

Table 8.6 Succession for urban ecosystems in Lima, own authorship

Succession for urban ecosystems in Lima			
Succession Stage/ Ecosystem	Pioneer	Seral	Climax
Wetland Ecosystem	Algae, phytoplankton and zooplankton	Giant reed (*Arundo donax*), cattail (*Typha sp*)	White heron, water lily
River Ecosystem	Giant reed (*Arundo donax*), Tamarisk (*Tamarix sp.*)	Elder (*Sambucus peruviana*), blue and white swallow (*Pygochelidon cyanoleuca peuviana*)	Humboldt willow (*Salix humboldtiana*), Peruvian pepper tree (*Schinus molle*), Peru coast toad (*Rhinella limensis*)
Industrial, contaminated ecosystem	Castor oil plant (*Ricinus communis*), Amaranth (*Amarathus dubius*)	Humboldt willow (*Salix humboldtiana*)	Peruvian pepper tree (*Schinus molle*), Palo verde (*Parkinsonia aculeata*)
Residential and garden ecosystem	Maidenhair (*Adiantum cuneatum*)	Introduced species	Introduced species
Densely inhabited ecosystems in historic Lima	Dandelion (*Traxacum officinale*), plantain (*Plantago major*)	Yellow trumpetbush (*Tecoma stans*)	Peruvian pepper tree (*Schinus molle*)
Park ecosystem	Kikuyu grass (*Pennisetum clandestinum*)	Yellow trumpetbush (*Tecoma stans*), introduced species	Introduced ornamental trees
Dry salty sandbar Ecosystem	Shoreline purslane (*Sesuvium portulacastrum*)	Seashore saltgrass (*Distichlis spicata*)	Palo Verde (*Parkinsonia aculeata*)
High transit pedestrian crossings/ avenues	Amaranth (*Amarathus dubius*)	Higuereta (*Ricinus communis*)	Peruvian pepper tree (*Schinus molle*)
Patio and rooftop ecosystems	Dandelion (*Traxacum officinale*), plantein (*Plantago major*)	Introduced species	Introduced species
Bio-orchards	Kikuyu (*Pennisetum clandestinum*)	Dandelion (*Traxacum officinale*), plantein (*Plantago major*)	Vegetable

soil, enabling germination of seeds or pieces of plants transported by the river. A lot of the species are neophytes. In rivers we can also find temporary islands that will be swiped the next time the river grows. These might have temporary pioneer species and birds that will use the island for nesting. In river basins in Lima we can find native species of fast growth like the Humboldt willow (*Salix humboldtiana*), the Peruvian pepper tree (*Schinus molle*), the Peruvian elder (*Sambucus peruviana*), that could be used for the re-naturalisation of river banks of the three basins that form Lima. There are also invasive species of ample distribution transported by the river such as the giant reed (*Arundo donax*) or the tamarisk (*Tamarix sp.*), introduced species from Australia in the 70 s into the National Reserve of Paracas to stop progress of the dunes. The main birds in the Rimac river are the Blue-and-white swallow (*Pygochelidon cyanoleuca peuviana*), the black vulture (*Coragyps atratus*) and the grey-headed gull (*Chroicocephalus cirrocephalus*) that

Table 8.7 Pioneer native species for the re-naturalization and re-conquest of natural spaces, own authorship

Pioneer native species for the re-naturalization and re-conquest of natural spaces	
Ecosystem	Species
Riverside	Elder (*Sambucus peruviana*), Peruvian Pepper tree (*Schinus molle*), Humboldt willow (*Salix humboldtiana*)
Cliffs	Shoreline purslane (*Sesuvium portulacastrum*)
Dunes and sandbanks	Shoreline purslane (*Sesuvium portulacastrum*), Palo Verde (*Parkinsonia aculeata*), yellow trumpetbush (*Tecoma stans*)
Wetlands	Giant reed (*Arundo donax*), cattail (*Typha sp*)
Hills	Amancae (*Ismene Amancaes*), old man's beard (*Tillandsia sp.*)

migrates when there is draught in the highlands. All these species will constitute the resilience of the ecosystem and can be used for its restoration and re-naturalization (see Tables 8.1 and 8.6).

Even though most original species of the basin have been displaced by introduced species, especially plants. In the case of the fauna, there are still original species. It is common to find, for example, the Peruvian pepper tree (*Schinus molle*) in abandoned or border ecosystems. With climate change the appearance of some species can be noted such as the Peruvian coast toad (*Rhinella limensis*), which has appeared in parks and gardens in the city due to an increase of rainfall in the last years.

As the city grows, the patches of vegetation become smaller until they are tiny and increase in number together with the number of constructed patches (Francis and Chadwick 2013: 45). Patch dynamic is constant (Luck and Wu 2002 cit. Francis and Chadwick 2013: 46). The return of ecosystems is slow and depends on the socio-economic processes in the city (Francis and Chadwick 2013: 46). This includes ecosystem adaptation and succession whose speed depends of the facilitation of ecological processes. Re-naturalisation ecology helps do this. It is important to note that according to the *Cliff urban hypothesis* developed by Larson et al. 2004, marginal spaces like precipices, cliffs and borders are essential in the reconquest of green spaces. Here, nature subsists. This is certainly true in cities like Berlin or New York. In spaces like Lima where it hardly rains, they would be located on riverbanks, on the slopes around the cities or in marginal zones between constructions. Some pioneer species could be key for the re-naturalisation of ecosystems in Lima (see Table 8.7).

References

CEPAL, FAO, IICA (2017) Comisión Económica para América Latina y El Caribe, Organización de las Naciones Unidas para la Agricultura y la alimentación, Instituto Interaméricano de Cooperación para la Agricultura. Perspectivas de la agricultura y del desarrollo rural en las Américas: una mirada hacia América Latina y El Caribe 2017–2018. San José, p 266 p

Francis R, Chadwick M (2013) Urban ecosystems. Understanding the Human Environment. Routledge, USA, p 220

Riley A (1998) "What Is Restauration?" from restoring strems in cities. In: Wheeler S, Beatley T (eds) (2014) The sustainable urban development, 3rd edn. Routledge, London and New York, pp 184–189

SERFOR (2019) Servicio Nacional Forestal y de Fauna Silvestre. Aves que conviven con los limeños. https://www.serfor.gob.pe/noticias/las-aves-silvestres-que-conviven-con-limenos. Revised: 17/12/19

Turner M (2001) Landscape ecology: in theory a practice pattern and process. Springer, New York, p 401p

Wohlgemuth T, Jentsch A, Rupert S (eds) (2019) Störungs-ökologie. Utb Haupt Verlag, Gernany, p 396

Zerbe S (2019) Renaturierung von Ökosystemen im Spannungsfeld von Mensch und Umelt. Springer Spektrum, Germany

Appendix A
Plant Species Index

Abstract The appendix shows the main species used in landscaping in Lima. The first part lists the main species classified according to way of growth in trees, fruit trees, palm trees, bushes, vines, indoor semi-perennials, outdoor semi-perennials, flowers and bulbs indicating its botanical classification, family and height. Subsequently, species are numbered according to their way of growth, organizing them according to characteristics of each species such as height, radicular depth, speed of growth, flower colour, blooming season in case of trees. Species are also listed based on size, flower colour, resistance to salt, sensibility to pests and disease as well as other characteristics useful for those that design landscapes in Lima city. In this sense, it is a practical appendix with characteristics appearing in user friendly charts which enable the design of green spaces.

Keywords Botanical change; Species; Trees; Bushes; Flowers

The appendix shows the main species used in landscaping in Lima. The first part lists the main species classified according to way of growth in trees, fruit trees, palm trees, bushes, vines, indoor semi-perennials, outdoor semi-perennials, flowers and bulbs indicating its botanical classification, family and height. Subsequently, species are numbered according to their way of growth, organizing them according to characteristics of each species such as height, radicular depth, speed of growth, flower colour, blooming season in case of trees. Species are also listed based on size, flower colour, resistance to salt, sensibility to pests and disease as well as other characteristics useful for those that design landscapes in Lima city. In this sense, it is a practical appendix with characteristics appearing in user friendly charts which enable the design of green spaces.

A.1 Most Used Plant Species in the Landscape of Lima City

The first part lists the main species classified according to their way of growth in: trees (see Table A.1), palm trees (see Table A.2), fruit trees (see Table A.3), bushes (see Table A.4), vines (see Table A.5), semi-perennial small plants (see Table A.6), indoor semi-perennials (see Table A.7), flowers (see Table A.8) and bulbs (see Table A.9) indicating botanical classification, family and height.

Tecoma stans known as Yellow Trumpet-shrub or Huaranguay, is a native species to the lower mountains in Peru that is distinctive in Lima landscape of this decade (see Sect. 3.2.1 and Fig. A.1).

A. Sabogal, *Urban Ecology*, Sustainable Development Goals Series,
https://doi.org/10.1007/978-3-030-69905-5

Table A.1 The most used tree species in Lima city

Trees		
Scientific name	Family	Height (m)
Araucaria araucana	Araucariaceae	30
Araucaria excelsa	Araucariaceae	30
Bauhinia aculeata	Fabaceae	30
Brugmancia arborea	Solanaceae	1.5
Callistemon citrinus	Myrtaceae	8
Casuarina equisetifolia	Casuarinaceae	25
Cedrela odorata	Meliaceae	20
Ceiba trichistandra	Bombacaceae	24
Cupressus funebris	Cupresaceae	6
Cupressus sempervivens	Cupresaceae	6
Delonix regia	Fabaceae	15
Dracaena draco	Asparagaceae	8
Eucalyptus camaldulensis	Myrtaceae	21
Eucalyptus globulus	Myrtaceae	21
Euphorbia pulcherrima	Euphorbiaceae	3
Ficus benjamina	Moraceae	18
Ficus elástica	Moraceae	28
Ficus pandurata	Moraceae	25
Fraxinus excelsior	Oleaceae	25
Ginkgo biloba	Ginkgophyta	25
Harpullia pendula	Sapindaceae	18
Jacaranda mimosifolia	Bignonaceae	18
Magnolia grandiflora	Magnoliaceae	30
Myoporum laetum	Myoporaceae	1.6
Parkinsonia aculeata	Fabaceae	12
Plumeria rubra	Apocynaceae	3.5
Populus deltoides	Salicales	20
Populus nigra	Salicales	18
Prosopis pallida	Fabaceae	30
Quercus robur	Fagaceae	20
Salix babylonica	Salicaceae	9
Salix humboldtiana	Salicaceae	25
Schefflera actinophylla	Araliaceae	26
Schinus molle	Anacardiaceae	8
Schinus terebinthifolius	Anacardiaceae	10
Spathodea campanulata	Bignoniaceae	16
Tecoma stans	Bignoniaceae	16
Thuja occidentalis	Cupressaceae	9
Tipuana tipu	Fabaceae	30

Fig. A.1 *Tecoma stans. Author* Ana Sabogal

Table A.2 The most used palm trees in Lima city

Palm trees		
Chrysalidocarpus lutescens	Arecaceae	8 m
Cocos nucifera	Arecaceae	12 m
Cycas revoluta	Cycadaceae	6 m
Hyophorbe lagenicaulis	Arecaceae	30 m
Phoenix dactylifera	Arecaceae	20 m
Roystonea regia	Arecaceae	25 m
Washingtonia filifera	Arecaceae	10 m

Table A.3 The most used fruit trees in Lima city

Fruit trees		
Carya illinoinensis	Juglandaceae	40 m
Cydonia oblonga	Rosaceae	8 m
Ficus carica	Moraceae	15 m
Inga feuilleei	Fabaceae	30 m
Juglans neotropica	Juglandaceae	40 m
Malus domestica	Rosaceae	8 m
Mangifera indica	Anacardiaceae	25 m
Olea europaea	Oleaceae	18 m
Persea americana	Lauraceae	30 m
Punica granatum	Punicaceae	1.5 m
Pyrus communis	Rosaceae	15 m

Table A.4 The most used bush species in Lima city

Bushes		
Abutilon pictum	Malvaceae	60 cm
Acalypha wilkesiana	Euphorbiaceae	70 cm
Breynia nivosa	Phyllanthaceae	40 cm
Bruncelsia calicina	Solanaceae	1.2 m
Caesalpinia pulcherrima	Fabaceae	5 m
Codiaeum variegatum	Euphorbiaceae	90 cm
Cordyline terminalis	Asparaceae	80 cm
Cupressus leylandii	Cupreaceae	60 cm
Cupressus macrocarpa	Cupresaceae	70 cm
Dieffenbachia sp.	Araceae	40 cm
Dracaena sp.	Asparagaceae	80 cm
Fuchsia coccinea	Onagraceae	50 cm
Heliconia sp.	Heliconiaceae	1.2 m
Heliotropium peruvianum	Braginaceae	75 cm
Hibiscus rosa-sinensis	Malvaceae	80 cm
Hydrangea macrophylla	Hydrangeaceae	80 cm
Ixora coccinea	Rubiaceae	60 cm
Lantana cámara	Verbenaceae	70 cm
Malvaviscus arboreus	Malvaceae	60 cm
Nandina domestica	Berberidaceae	1.2 m
Nerium oleander	Apocinaceae	90 cm
Plumbago capensis	Plumbaginaceae	70 cm
Punica granatum	Punicaceae	60 cm
Ravenala madagascariensis	Strelitziaceae	3 m
Sambucus peruviana	Adoxaceae	1.5 m
Sanchezia nobilis	Acanthaceae	1.4 m
Schefflera arborícola	Araliaceae	70 cm
Spartium junceum	Leguminosaceae	90 cm
Streptosolen jamesonii	Solanaceae	50 cm
Yucca elephantipes	Asparagaceae	1 m

A.2 Use of the Plant Species According to the Space

In this part species are numbered according to their way of growth, organizing them according to the characteristics of each species such as: size (see Tables A.10, A.12, A.15 abd A.25), radicular depth (see Table A.12), speed of growth (see Tables A.11 and A.16), flower colour (see Tables A.13, A.21 and A.26), blooming season (see Table A.18), resistance to salt (see point A.2.6), sensibility to pests and disease (see Table A.23) as well as other characteristics useful for those that design landscapes in Lima city. In this sense, it is a practical part of the appendix with characteristics appearing in user friendly charts which enable the design of green spaces.

A.2.1 Trees

In this section tree species are numbered according to their way of growth, organizing them according to characteristics of each species such as height (see Table A.10),

Table A.5 The most used vines in Lima city

Vines		
Allamanda cathartica	Apocinaceae	1 m
Antagonon leptopus	Poligonaceae	60 cm
Bougainvillea spectabilis	Nistaginaceae	1.2 m
Clerodendrum speciosum	Verbenaceae	1.2 m
Epipremnum aureum	Araceae	60 cm
Hedera hélix	Araliaceae	60 cm
Ipomoea violácea	Convolvulaceae	35 cm
Jasminum humile	Oleaceae	1.4 m
Jasminum officinale	Oleaceae	1.4 m
Jasminum polyanthum	Oleaceae	1.2 m
Lathyrus odoratus	Leguminosaceae	90 cm
Lonicera periclymenum	Caprifoliaceae	90 cm
Monstera deliciosa	Araceae	1.20 m
Pelargonium petate	Geraniaceae	60 cm
Pyrostegia ígnea	Bignoniaceae	1.2 m
Stephanotis floribunda	Apocynaceae	1.20 m
Tropaeolum majus	Tropaeolaceae	1.2 m

Table A.6 The most used semi-perennial plant species in Lima city

Semi-perennial, small plants		
Alternanthera reineckii	Amaranthaceae	30 cm
Asparagus densiflorus	Asparagaceae	40 cm
Asparagus sprengeri	Asparrgaceae	40 cm
Beloperone guttata	Acanthaceae	40 cm
Canna indica	Cannaceae	60 cm
Chlorophytum comosum	Liliaceae	30 cm
Coleus blumei	Lamiaceae	30 cm
Colocasia sp.	Araceae	90 cm
Cuphea hyssopiolia	Lythraceae	15 cm
Distichlis spicata	Poaceae	15 cm
Fuchsia coccinea	Onagraceae	30 cm
Gardenia jasminoides	Caprifoliaceae	70 cm
Gerbera jamesonii	Compositae	25 cm
Leucanthemum vulgare	Compositae	30 cm
Mesembryanthemum spectabile	Aisoaceae	25 cm
Oxalis sp.	Oxalidaceae	20 cm
Pachystachys lutea	Acantaceae	30 cm
Pelargonium hortorum	Geraniaceae	60 cm
Plectrathus verticillatus	Lamiaceae	10 cm
Strelitzia reginae	Strelitziaceae	80 cm
Verbena peruviana	Verbenaceae	30 cm

Table A.7 The most used semi-perennial indoor species in Lima city

Semi-perennial indoor plants		
Adiantum cuneatum	Pteridaceae	25 cm
Anthurium andreanum	Araceae	60 cm
Begonia masoniana	Begnoniaceae	35 cm
Begonia rex x cultorum	Begnoniaceae	30 cm
Calathea sp.	Maranthaceae	30 cm
Gloxinia perennis	Gesneriaceae	25 cm
Saintpaulia ionantha	Gesneriaceae	15 cm
Spathiphyllum wallisii	Araceae	70 cm

Table A.8 The most used flower species in Lima city

Flowers		
Antirrhinum majus	Plantaginaceae	30 cm
Aster sp.	Asteraceae	25 cm
Begonia semperflorens	Begoniaceae	20 cm
Catharanthus roseus	Apocynaceae	30 cm
Chrysanthemum morifolium	Compositae	30 cm
Dianthus caryophyllus	Cariofilaceae	20 cm
Gerbera jamesonii	Geraniaceae	60 cm
Impatiens balsamina	Balsaminiaceae	20 cm
Matthiola incana	Crusiferaceae	25 cm
Petunia x hybrida	Solanaceae	30 cm
Phlox drummondii	Polemoniaceae	15 cm
Salvia splendens	Labiaceae	25 cm
Tagetes patula	Compositae	30 cm
Viola x wittrockiana	Violaceae	15 cm
Zinnia elegans	Asteraceae	25 cm

Table A.9 The most used bulbs species in Lima city

Bulbs		
Agapanthus africanus	Amaryllidaceae	60 cm
Eucharis x grandiflora	Amaryllidiaceae	50 cm
Hippeastrum miniatum	Amaryllidaceae	50 cm
Iris germánica	Iridaceae	40 cm
Ismene Amancaes	Amaryllidaceae	45 cm
Lirium longuiflorum	Liliaceae	60 cm
Narcissus pseudonarcissus	Amaryllidaceae	45 cm
Polianthes tuberosa	Apsaragaceae	80 cm
Tulipa gesneriana	Liliaceae	45 cm

Table A.10 Tree classification according to height

Tree classification according to height		
Tall trees (more than 20 m.)	Medium-sized trees (10–20 m.)	Small trees (under 10 m.)
Araucaria araucana	*Acasia macracantha*	*Brugmancia arborea*
Ceiba trichistandra	*Bauhinia aculeata*	*Callistemon citrinus*
Eucalyptus globulus	*Delonix regia*	*Cupressus funebris*
Ficus pandurata	*Ficus benjamina*	*Cycas revoluta*
Fraxinus excelsior	*Ginkgo biloba*	*Dracaena draco*
Juglans neotropica	*Harpullia pendula*	*Euphorbia pulcherrima*
Magnolia grandiflora	*Inga feuilleei*	*Ficus carica*
Populus nigra	*Jacaranda mimosifolia*	*Myoporum laetum*
Prosopis pallida	*Olea europarea*	*Salix babylonica*
Roystonea regia	*Parkinsonia aculeata*	*Schefflera arboricola*
Schefflera actinophylla	*Schinus terebinthifolius*	*Schinus molle*
Tipuana tipu	*Washingtonia filifera*	*Thuja occidentalis*

Fig. A.2 *Myoporum laetum. Author* Ana Sabogal

speed of growth (see Table A.11), radicular depth (see Table A.12), flower and leaf colour (see Table A.13) and soil quality (see Table A.14) in the case of trees (Figs. A.2 and A.3).

A.2.2 Shrubs

In this section, species are numbered according to their way of growth, organizing them according to the characteristics of each species

Table A.11 Tree classification according to speed of growth

Tree classification according to speed of growth	
Fast growing trees	Slow growing trees
Cedrela odorata	*Araucaria araucana*
Ceiba trichistandra	*Cupressus funebris*
Eucaliptus camadulensis	*Cycas revoluta*
Ficus elastica	*Delonix regia*
Juglans neotropica	*Dracaena draco*
Parkinsonia aculeata	*Ginkgo biloba*
Salix babylonica	*Magnolia grandiflora*
Salix humboldtiana	*Quercus robur*
Tipuana tipu	*Thuja occidentalis*

Table A.12 Tree classification according to radicular depth

Tree classification according to radicular depth		
Superficial root size (0.5–2 m.)	Middle root size (2–3 m.)	Deep size (3–4.5 m.)
Callistemon citrinus	*Ginkgo biloba*	*Casuarina equisetifolia*
Cycas revoluta	*Bauhinia aculeata*	*Ceiba trichistandra*
Dracaena draco	*Jacaranda mimosifolia*	*Eucalyptus globulus*
Euphorbia pulcherrima	*Persea amecicana*	*Ficus elastica*
Roystonea regia	*Salix babylonica*	*Populus nigra*
Schinus molle	*Schinus terebinthifolius*	*Prosopis pallida*

*It exists a directly relationship between the tree size and the root deep

such as: size (see Table A.15), speed of growth (see Table A.16), light necessities (see Table A.17), blooming season (see Table A.18) and soil necessities (see Table A.19) (Fig. A.4).

A.2.3 Vines

In this section species are numbered according to: their speed of growth (see Table A.20) and flower colour (see Table A.21) (Figs. A.5 and A.6).

A.2.4 Semi-Perennial Plants

In this section species are numbered according to their light necessities (see Table A.22), soil requirement and plague and disease sensibilities (see Table A.23) and their leaf texture (see Table A.24) (Fig. A.7).

Fig. A.3 *Schinus molle*. *Author* Ana Sabogal

Table A.13 Tree classification according to flower and leaf color

Tree classification according to flower and leaf color		
Flower color	Leaf color	
	Dark green	Light green
Pink: *Ceiba trichistandra*	*Araucaria excelsa*	*Ficus benjamina*
Lilac: *Jacaranda mimosifolia*	*Cupressus funebris*	*Ginkgo biloba*
White: *Magnolia grandiflora, Brugmancia arborea*	*Ficus elastica*	*Parkinsonia aculeata*
Red: *Callistemon citrinus,* *Euphorbia pulcherrima*	*Jacaranda mimosifolia*	*Populus deltoides*
Yellow: *Acasia macracantha,* *Parkinsonia aculeata*	*Quercus robur*	*Salix babylonica*
Brown fruit: *Fraxinus excelsior*	*Thuja occidentalis*	*Schefflera actinophylla*

Table A.14 Tree classification according to soil quality

Tree classification according to soil quality	
Trees without soil quality demands and little water necessities	Trees with soil quality demands
Casuarina equisetifolia	*Bauhinia aculeata*
Eucaliptus camadulensis	*Cupressus funebris*
Ficus carica	*Cycas revoluta*
Ficus elástica	*Fraxinus excelsior*
Harpullia pendula	*Ginkgo biloba*
Parkinsonia aculeata	*Magnolia grandiflora*
Prosopis pallida	*Thuja occidentalis*

Table A.15 Shrub classification according to height characteristics

Shrub classification according to height characteristics		
Small sized shrubs	Middle sized shrubs	Tall sized shrubs
Breynia nivosa	*Acalypha wilkesiana*	*Brugmancia arborea*
Hydrangea macrophylla	*Bruncelsia calicina*	*Caesalpinia pulcherrima*
Ixora coccinea	*Codiaeum variegatum*	*Cofea arabica*
Lantana cámara	*Dracaena sp.*	*Nerium oleander*
Punica granatum	*Malvaviscus arboreus*	*Plumeria rubra*
Schefflera arborícola	*Nandina domestica*	*Schefflera arboricola*
Streptosolen jamesonii	*Yucca elephantipes*	*Spartium junceum*

Table A.16 Shrub classification according to speed of growth

Shrub classification according to speed of growth	
Fast growing shrubs	Slow growing shrubs
Abutilon pictum	*Breynia nivosa*
Caesalpinia pulcherrima	*Bruncelsia calicina*
Ixora coccinea	*Codiaeum variegatum*
Lantana cámara	*Cordyline terminalis*
Malvaviscus arboreus	*Cupressus leylandii*
Nerium oleander	*Hydrangea macrophylla*
Sanchezia nobilis	*Nandina domestica*
Schefflera arboricola	*Sambucus peruviana*
Spartium junceum	*Yucca elephantipes*

Table A.17 Shrub classification according to light necessities

Shrub classification according to light necessities	
Shrubs with huge light necessities	Shrubs with little light necessities
Acalypha wilkesiana	*Abutilon pictum*
Bruncelsia calicina	*Breynia nivosa*
Heliotropium peruvianum	*Cordyline terminalis*
Hibiscus rosa-sinensis	*Dieffenbachia sp.*
Malvaviscus arboreus	*Dracaena sp.*
Nerium oleander	*Fuchsia coccinea*
Plumbago capensis	*Hydrangea macrophylla*
Ravenala madagascariensis	*Ixora coccinea*
Sambucus peruviana	*Sanchezia nobilis*
Spartium junceum	*Streptosolen jamesonii*

Table A.18 Shrub classification according to soil necessities

Shrub classification according to soil necessities	
Shrubs without soil quality demands	Rich soil necessities
Acalypha wilkesiana	*Abutilon pictum*
Cupressus macrocarpa	*Breynia nivosa*
Heliconia sp.	*Bruncelsia calicina*
Heliotropium peruvianum	*Codiaeum variegatum*
Hibiscus rosa-sinensis	*Cordyline terminalis*
Lantana cámara	*Dieffenbachia sp.*
Malvaviscus arboreus	*Dracaena sp.*
Nerium oleander	*Fuchsia coccinea*
Punica granatum	*Hydrangea macrophylla*
Sambucus peruviana	*Ixora coccinea*
Spartium junceum	*Nandina domestica*

Fig. A.4 *Hibiscus rosa-sinensis*. *Author* Ana Sabogal

Table A.19 Shrub classification according to soil necessities

Shrub classification according to blooming season		
Blooming the whole year	Blooming in spring	Blooming in summer
Fuchsia coccinea	*Gardenia jasminoides*	*Abutilon pictum*
Hibiscus rosa-sinensis	*Malvaviscus arboreus*	*Bruncelsia calicina*
Ixora coccinea	*Sambucus peruviana*	*Caesalpinia pulcherrima*
Lantana cámara	*Bruncelsia calicina*	*Hydrangea macrophylla*
Plumbago capensis	*Nandina domestica*	*Nerium oleander*
Spartium junceum	*Sanchezia nobilis*	*Punica granatum*
Streptosolen jamesonii	*Sambucus peruviana*	*Ravenala madagascariensis*

Table A.20 Vine classification according to speed of growth

Vine classification according to speed of growth	
Fast growing vines	Slow growing vines
Allamanda cathartica	*Bougainvillea spectabilis*
Antagonon leptopus	*Hedera hélix*
Epipremnum aureum	*Jasminum humile*
Ipomoea violácea	*Jasminum polyanthum*
Lathyrus odoratus	*Lonicera periclymenum*
Tropaeolum majus	*Pyrostegia ígnea*

Fig. A.5 Vines *Tropaeolum majus* used as a garden bed. *Author* Ana Sabogal

Table A.21 Vine classification according to flower color

Vine classification according to flower color	
Scientific name	Flower color
Allamanda cathartica	Yellow
Antagonon leptopus	White or pink
Bougainvillea spectabilis	Fuchsia, rot, orange, white or pink
Clerodendrum speciosum	Rot with blue and white
Ipomoea violácea	Violet
Jasminum humile	Yellow
Jasminum officinale	White
Jasminum polyanthum	White
Lathyrus odoratus	Violet, blue, rot or white
Lonicera periclymenum	Yellow or white
Pelargonium petate	Pink or white
Pyrostegia ígnea	Orange
Stephanotis floribunda	White
Tropaeolum majus	Orange

A.2.5 Blooming Plants, Bulbs and Indoor Semi-Perennial Small Plants

In this section species are numbered according to their size (see Table A.25) and flower colour (see Table A.26) (Fig. A.8).

A.2.6 Salt Tolerant Plant Used in the Landscape from Lima

The importance of selecting plants that resist salt brises is cardinal for Lima city (see Sect. 3.2.1). In this point plant species that tolerate marine breeze are listed. Semi-perennial indoor species, flowers and bulbs are not resistant to soil salinity and marine breeze (see Table A.27).

Fig. A.6 *Bougainvillea spectabilis. Author* Ana Sabogal

Table A.22 Semi-perennial plant classification according to their light necessities

Semi-perennial plant classification according to their light necessities	
Direct light	Indirect light
Acalypha wilkesiana	*Anthurium andreanum*
Chlorophytum comosum	*Breynia nivosa*
Codiaeum variegatum	*Beloperone guttata*
Coleus blumei	*Colocasia sp.*
Mesembryanthemum spectabile	*Eucharis x grandiflora*
Pachystachys lutea	*Fuchsia coccinea*
Pelargonium hortorum	*Hydrangea macrophylla*
Salvia splendens	*Ixora coccinea*

Table A.23 Semi-perennial plant classification according to their soil requirement and plague and disease sensibilities

Semi-perennial plant classification according to their soil requirement and plague and disease sensibilities	
Low quality soil requirements and low plague and disease sensibilities	High soil requirement and high plague and disease sensibilities
Althernanthera reineckii	*Anthurium andreanum*
Canna indica	*Begonia masoniana*
Chlorophytum comosum	*Beloperone guttata*
Coleus blumei	*Breynia nivosa*
Cuphea hyssopiolia	*Eucharis x grandiflora*
Lantana cámara	*Fuchsia coccinea*
Mesembryanthemum spectabile	*Hydrangea macrophylla*
Pelargonium hortorum	*Leucanthemum vulgare*
Plectrathus verticillatus	*Pachystachys lutea*
Strelitzia reginae	*Verbena peruviana*

Table A.24 Semi-perennial plants classification according to their leaf texture

Semi-perennial plants classification according to their leaf texture		
Depth	Middle	Fine
Begonia masoniana	*Begonia rex x cultorum*	*Adiantum cuneatum*
Calathea sp.	*Beloperone guttata*	*Althernanthera reineckii*
Canna indica	*Catharanthus roseus*	*Asparagus densiflorus*
Hydrangea macrophylla	*Chlorophytum comosum*	*Cuphea hyssopiolia*
Ixora coccinea	*Coleus blumei*	*Impatiens balsamina*
Mesembryanthemum spectabile	*Fuchsia coccinea*	*Oxalis sp.*
Pelargonium *hortorum*	*Pachystachys lutea*	*Plectrathus verticillatus*
Spathiphyllum wallisii	*Salvia splendens*	*Verbena peruviana*

Fig. A.7 *Mesembryanthemum spectabile* small semi-perennial plant. *Author* Ana Sabogal

Table A.25 Classification of blooming plants, bulbs and semi-perennial small plants according to their size

Classification of blooming plants, bulbs and semi-perennial small plants according to their size		
Small (10–30 cm.)	Middle (30–60 cm.)	Big (more than 60 cm.)
Adiantum cuneatum	*Antirrhinum majus*	*Agapanthus africanus*
Aster sp.	*Beloperone guttata*	*Anthurium andreanum*
Begonia rex x cultorum	*Catharanthus roseus*	*Canna indica*
Begonia semperflorens	*Chrysanthemum morifolium*	*Colocasia sp.*
Dianthus caryophyllus	*Eucharis x grandiflora*	*Gardenia jasminoides*
Gloxinia perennis	*Gerbera jamesonii*	*Heliotropium peruvianum*
Impatiens balsamina	*Hippeastrum miniatum*	*Hydrangea macrophylla*
Mathiola incana	*Iris germánica*	Ixora (*Ixora coccinea*)
Phlox drummondii	*Ismene Amancaes*	Lirio (*Lirium longuiflorum*)
Saintpaulia ionantha	*Leucanthemum vulgare*	*Pelargonium hortorum*
Salvia splendens	*Narcissus pseudonarcissus*	*Polianthes tuberosa*
Tagetes patula	*Petunia x hybrida*	*Spathiphyllum wallisii*
Viola x wittrockiana	*Zinnia elegans*	*Strelitzia reginae*

Table A.26 Classification of blooming plants, bulbs and semi-perennial small plants according to their flower color

Classification of blooming plants, bulbs and semi-perennial small plants according to their flower color	
Agapanthus africanus	Violet
Anthurium andreanum	Rot
Antirrhinum majus	White, pink, rot, yellow, violet
Aster sp.	Violeta
Begonia semperflorens	White or pink
Catharanthus roseus	White, pink, rot, violet
Chrysanthemum morifolium	White, yellow, violet
Dianthus caryophyllus	White, pink, rot
Eucharis x grandiflora	White
Gerbera jamesonii	White, rot, orange
Gloxinia perennis	White, rot, violet
Hippeastrum miniatum	White, rot and white with rot
Impatiens balsamina	White, orange, rot
Iris germánica	White, violet
Ismene Amancaes	Yellow
Lirium longuiflorum	White
Mathiola incana	White, rosado, violet
Narcissus pseudonarcissus	White, yellow
Petunia x hybrida	White, pink, violet
Phlox drummondii	White, rot, violet
Polianthes tuberosa	White
Saintpaulia ionantha	White, pink, violet
Salvia splendens	Rot
Spathiphyllum wallisii	White
Tagetes patula	Orange
Tulipa gesneriana	White, pink, rot, yellow
Viola x wittrockiana	White, yellow and violet combined
Zinnia elegans	White, rosado, yellow, orange

Fig. A.8 *Antirrhinum majus*. *Author* Ana Sabogal

Table A.27 Plant species that tolerate soil salinity and marine breeze, own authorship

Plant species that tolerate soil salinity and marine breeze			
Trees	Shrubs	Vine	Semi-perennial, small plants
Acasia macracantha	*Abutilon pictum*	*Bougainvillea spectabilis*	*Alternanthera reineckii*
Araucaria Araucana	*Acalypha wilkesiana*	*Ipomoea violacea*	*Chlorophytum comosum*
Casuarina equisetifolia	*Caesalpinia pulcherrima*	*Lonicera periclymenum*	*Coleus blumei*
Dracaena draco	*Hibiscus rosa-sinensis*	*Pyrostegia ígnea*	*Cuphea hyssopiolia*
Ficus carica	*Lantana cámara*		*Mesembryanthemum spectabile*
Mioporum laetum	*Malvaviscus arboreus*		
Parkinsonia aculeata	*Nerium oleander*		
Prosopis pallida	*Punica granatum*		
Tipuana tipu	*Spartium junceum*		